CHEMISTRY RESEARCH AND APPLICATIONS

IRIDIUM

OCCURRENCE, CHARACTERISTICS AND APPLICATIONS

Chemistry Research and Applications

Additional books in this series can be found on Nova's website
under the Series tab.

Additional E-books in this series can be found on Nova's website
under the E-book tab.

Materials Science and Technologies

Additional books in this series can be found on Nova's website
under the Series tab.

Additional E-books in this series can be found on Nova's website
under the E-book tab.

CHEMISTRY RESEARCH AND APPLICATIONS

IRIDIUM

OCCURRENCE, CHARACTERISTICS AND APPLICATIONS

CHIEKO FUKUI
AND
MAEMI ONO
EDITORS

Nova Science Publishers, Inc.
New York

For permission to use material from this book please contact us:
Telephone 631-231-7269; Fax 631-231-8175
Web Site: http://www.novapublishers.com

Library of Congress Cataloging-in-Publication Data

Iridium : occurrence, characteristics, and applications / [edited by] Chieko Fukui and Maemi Ono.
p. cm.
Includes bibliographical references and index.
ISBN 978-1-62081-681-3 (hardcover)
1. Iridium. I. Fukui, Chieko. II. Ono, Maemi.
QD181.I7I75 2011
546'.643--dc23
2012013596

Published by Nova Science Publishers, Inc. † New York

CONTENTS

Chapter 4- Organometallic complexes of iridium have gained much importance because of their novelty, reactivity and applications in diverse fields, for example, the commercial homogeneous catalytic process called *'Cativa'* {Iridium (I) catalyst with Ru-complex activator} for carbonylation of methanol to acetic acid. Iridium carbonyl complexes of nitrogen-oxygen (N-O) and phosphorous-oxygen (P-O) donor ligands are of much interest in recent time. The N-O type ligands (L) like methyl picolinate (*a*), ethyl picolinate (*b*), methyl nicotinate (*c*), ethyl nicotinate (*d*), methyl isonicotinate (*e*) and ethyl isonicotinate (*f*) affords the tetra-coordinated neutral dicarbonyl complexes of the type $[Ir(CO)_2ClL]$. The square planar geometry with Ir in the central position and the intermolecular C–H···O and Ir···C interactions in the complex containing ethyl nicotinate greatly stabilize the supramolecular structure in the solid state. The steric, electronic and kinetics of oxidative addition (OA) of the complexes with electrophiles like CH_3I, C_2H_5I, I_2 etc. are interesting. The complexes, *cis*-$[Ir(CO)_2L_2](PF_6)$ (L = pyridine based ligands) showed Water Gas Shift Reaction (WGSR) activity. The Vaska's complex, *trans*-$[Ir(CO)Cl(PPh_3)_2]$, occupies a central position of pivotal importance in the discovery of OA reactions and several such complexes of ligands, $PMePh_2$, PMe_2Ph, PMe_3 were investigated. The P-O type ligands (L′) like 2-$Ph_2PC_6H_4COOMe$ (*a′*) and 2-$Ph_2PC_6H_4OMe$ (*b′*) afford tetra-coordinated mono-carbonyl Vaska's type complexes *trans*-$[Ir(CO)ClL'_2]$. The IR spectra of such complexes in $CHCl_3$ solution show the terminal υ (CO) bands at around 1957 and 1959 cm^{-1} respectively, which are significantly lower in frequency compared to Vaska's complex, *trans*-$[Ir(CO)Cl(PPh_3)_2]$ (1965 cm^{-1}) and therefore substantiate the enhanced electron density at the metal centre. The molecular structure of the complex containing 2-$Ph_2PC_6H_4COOMe$ ligand indicates iridium–oxygen (ester group) distances [Ir···O (2) 3.24 Å, Ir···O (5) 3.29 Å] and angle [O (5)···Ir··O (2) 157.25°] suggesting a long-range intramolecular "*Secondary*" Ir···O interaction resulting in a *pseudo*-hexacoordinated complex. The complexes undergo OA with small molecules like CH_3I, I_2 and O_2 to produce Ir (III) carbonyl species of the type $[Ir(CO)Cl(CH_3)IL'_2]$, $[Ir(CO)ClI_2L'_2]$ and $[Ir(O_2)(CO)\ Cl(2\text{-}Ph_2PC_6H_4OMe)_2]$ respectively. Some interesting mixed-ligand complex, $[IrCl(CO)(P^iPr_3)(Sb^iPr_3)]$, was also reported. Other complexes $[IrX(CO)_2(R_2M)]$, where, X = Cl, Br, I; M = P, As, Sb; R=C_1-C_{20} alkyl, alkoxy radicals may find application as hydrocarboxylation catalysts. Several water soluble complexes like $[Ir(CO)(TPPMS)_3]CF_3SO_3$, (TPPMS = $PPh_2(2\text{-}C_6H_4SO_3K)$, and their reactivities with small molecules (H_2 or CO) in polar solvents (DMSO or H_2O) were examined. The complex, $[Ir(COD)(PR_3)Py]PF_6$, where COD = 1,5-cyclooctadiene, PR_3 = tris-cyclohexylphosphine, and Py = pyridine, known as *Crabtree's* catalyst is active for homogeneous hydrogenation.

Chapter 5- An important variant for transition metal-catalyzed carbon-carbon or carbon-heteroatom bond formation is allylic substitution. The regio- and stereoselectivities of reactions using iridium catalysts are quite different from those of palladium-catalyzed reactions. Particularly, asymmetric synthesis of branched products based on the regioselective iridium-catalyzed allylic substitution of unsymmetrical substrates has been a subject of current interest. This review highlights the recent studies on iridium-catalyzed allylic alkylation, amination, etherification, and so on. In this review, the enantiocontrol in allylic substitutions using chiral iridium complexes is summarized.

Chapter 6- The use of iridium (Ir) as a material for amperometric electrodes, especially in biosensors, has been reported in several studies [1-5]. Research results have also been reported on the use of other materials [6], including binary alloys [7], for this purpose. However, using such materials is associated with inadequate hydrogen peroxide detection, susceptibility to interference species (e.g., ascorbic acid, uric acid, and acetaminophen), a high base current, and increased complexity in fabricating electrodes, among other problems [1-5]. Hydrogen peroxide is produced by oxidases, and the interference species exist in body fluids. Adequate hydrogen peroxide detection and insusceptibility to interference species are needed in amperometric biosensors. Accordingly, Platinum has almost always been used in the materials of amperometric biosensor electrodes; carbon is an exception [8-12]. Certain binary Ir alloys, i.e., Ir-Nb, Ir-Zr, and Ir-Nb-Zr, have been noted as having high temperature materials [13, 14]. However, the prices of rare metals including Ir have been fluctuating widely because of the global economic crisis and/or because they are mined in limited quantities. This has created a need to use alternative materials in place of Ir in various fields, including biosensor electrode fabrication. In our work, we found that an Ir-23 at.%Nb (Ir-23Nb) binary alloy can oxidize hydrogen peroxide better than any other Ir-Nb binary alloy.

Chapter 7- Since the seminal work on electrophosphorescent organic light-emitting diodes (PHOLEDs) based on fac-[Ir(ppy)3] (Hppy = 2-phenylpyridine), considerable effort has been devoted to the molecular design and synthesis of efficient phosphorescent heavy metal complexes in the fabrication of high-efficiency PHOLEDs, owing to their potential to harness the energies of both the singlet and triplet excitons in the electroluminescence process. Although the emission efficiency and the color of iridium-based PHOLEDs can be readily tuned by structural modifications of the ligand chromophores, the performance of PHOLEDs is largely governed by the charge balance between the electrons and holes from opposite electrodes. It is vital to develop electrophosphors that can show improved charge-balancing features and permits a complete energy transfer between the host and the dopant in the device.

The integration of hole-transporting carbazole module and phosphorescent iridium(III) cyclometalate core appears to be a good choice for the design of more efficient devices. We describe here the synthesis and device performance of a series of interesting small-molecule cyclometalated iridium(III) complexes functionalized with various carbazole units.

In: Iridium: Occurrence, Characteristics and Applications ISBN: 978-1- 62081-681-3
Editors: Chieko Fukui and Maemi Ono

Chapter 1

CHEMISTRY AND APPLICATIONS OF VOLATILE METAL-ORGANIC IRIDIUM COMPOUNDS

N. B. Morozova**, *N. V. Gelfond and I. K. Igumenov
Nikolaev Institute of Inorganic Chemistry, Siberian Branch of RAS, Novosibirsk, Russia

ABBREVIATIONS

XRD	X-ray diffraction analysis
Et	ethyl
acac	$[CH_3\text{-}CO\text{-}CH\text{-}CO\text{-}CH_3]^-$ – acetylacetonate anion
tfac	$[CF_3\text{-}CO\text{-}CH\text{-}CO\text{-}CH_3]^-$ – trifluoroacetylacetonate anion
hfac	$[CF_3\text{-}CO\text{-}CH\text{-}CO\text{-}CF_3]^-$ – hexafluoroacetylacetonate anion
thd	$[C(CH_3)_3\text{-}CO\text{-}CH\text{-}CO\text{-}C(CH_3)_3]^-$ – (2,2,6,6-tetramethylheptane-3,5-dionate) anion
dtc	$[S_2CNEt_2]^-$ – diethyldithiocarbamate anion
cod	C_8H_{12} – 1,5-cyclooctadiene
allyl	$[C_3H_5]^-$ – allyl anion
Cp	$[C_5H_5]^-$ – cyclopentadienyl anion
Cp'	$[CH_3\text{-}C_5H_4]^-$ – methylcyclopentadienyl anion
Cp*	$[(CH_3)_5\text{-}C_5]^-$ – pentamethylcyclopentadienyl anion
DTA	differential thermo analysis

INTRODUCTION

Metal iridium exhibits a rare combination of chemical and physical properties. Iridium is a representative of platinum group metals. It is stable against various chemical actions. In the

* E-mail: mor@niic.nsc.ru.

metal form, iridium is insoluble in mineral acids (including aqua regia), does not interact at a high temperature with the melts of other metals or silicates, does not form carbides when in contact with carbon, and possesses high resistance against oxidation. Iridium also has a high melting point (2443°C), and it is the only metal possessing unique mechanical characteristics in the air at temperatures above 1600°C.

Iridium-containing film materials found a broad application in protecting refractory materials in high-temperature processes of glass melting or crystal growth, in rocket techniques for the protection of combustion chambers in engines and turbine blades, in electrochemical processes to protect electrodes, in various areas of catalysis, in developing power optics elements for high-power lasers, for manufacturing film thermocouples, electric contacts, etc. The known methods of the deposition of iridium-containing layers include: physical deposition from the gas phase – thermal or electron beam evaporation of iridium in high vacuum followed by condensation, cathode spraying, magnetron sputtering, electrolysis of salt melts. During the recent decades, Chemical Vapour Deposition has become the most widespread method for depositing iridium coatings (MOCVD – Metal-Organic Chemical Vapour Deposition) [1-34]. This is one of the most dynamically developing methods of forming metal layers for different purposes. The deposition of films and coatings includes several main stages: the transfer of the precursor into the gas phase, vapour transport to the heated substrate, decomposition of vapour on the substrate surface with the formation of coating material and removal of gaseous reaction products. A distinguishing feature of the method is efficient control of the process parameters and therefore the required characteristics of the resulting layers and structures: crystallite size, thickness of the coating, its structure and morphology. Chemistry of precursors plays the key role in these processes because deposition parameters and the design of MOCVD installations are determined by the set of physicochemical characteristics of the compounds used. The major requirements to MOCVD precursors are volatility (ability to pass into the gas phase without decomposition at moderate temperatures (below 250°C)), thermal stability in the condensed and gas phases. As a rule, various classes of volatile coordination and organometallic compounds are used as MOCVD precursors. The active development of the chemistry of these compounds is connected with the great diversity of properties depending on the chemical composition and structure of complexes and, as a consequence, with the possibility of their practical application. There is a variety of the classes of volatile iridium-containing compounds – halides, homo- and heteroligand carbonyls, dienes, olefins, phosphines, allyls, amidinates, chelate complexes with {O,O} coordination centre (beta-diketonates), with {S,S} coordination centre (dithiocarbamate) and mixed (heteroligand) compounds. At present, almost all the usual types of the iridium organic compounds with ligands are known, the lignads being donors of one to six electrons. This chapter presents mainly the results obtained by the authors, with the necessary literature data involved. Below we will consider the synthetic methods of a number of different classes of volatile organometallic and coordination iridium compounds.

SYNTHESIS OF VOLATILE ORGANOMETALLIC AND COORDINATION IRIDIUM COMPOUNDS

Homo- and Heteroligand Iridium Carbonyls

Iridium carbonyls are obtained at high CO pressure. The representatives of these compounds are yellowish-green $Ir_2(CO)_8$ (sublimates) and bright-yellow $Ir_4(CO)_{12}$ (decomposed during heating). This is a number of heteroligand iridium carbonyls. Carbonylhalides are the starting compounds for the synthesis of substituted carbonylhalides. The carbonyl iridium complexes containing bidentate singly charged anions with O-donor atoms are synthesized using the reaction of $Na_2(Ir_2(CO)_4X_5)$ (X is halogen) with β-diketones in the presence of bases, or through the interaction of the derivatives of Ir(I), such as $Ir(CO)_2LCl$ (L is, for example, isopropyl amine) with acetylacetone according to the reaction described in [35]:

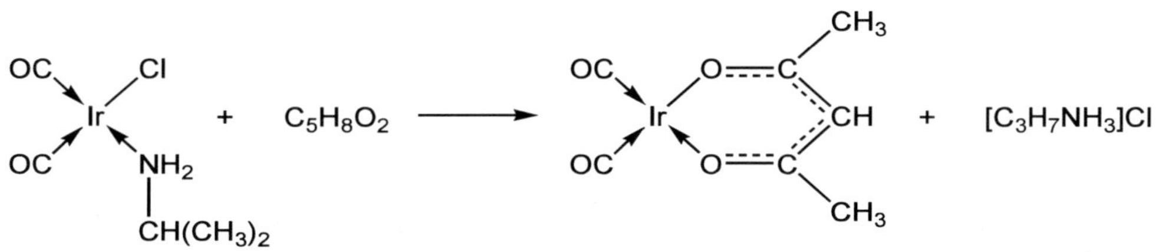

Dicarbonyl β-diketonates of Ir(I) are crystalline compounds of different colours (from green to yellow), stable in the air. The complexes sublime in vacuum at moderate temperatures [35]. $Ir(CO)_2(tfac)$ reacts with triphenylphosphine to form air-unstable yellow crystals of penta-coordinated $Ir(CO)(PPh_3)(tfac)$ [35]. $Ir(CO)_2(acac)$ is also obtained by passing CO through the hexane solution of $Ir(acac)(L)_2$, where L is ethylene, propylene, vinyl chloride, vinyl acetate etc. [36].

Through the interaction of methyl lithium and the tetrahydrothiophen (THT) complex $IrC1_3(THT)_3$ in diethyl ether and in the presence of tetramethylethylenediamine, the synthesis and spectral characterization of compound $[Li(TMED)]_3[IrMe_6]$ was carried out. This compound is unstable in the air and under heating. This complex was used for further synthesis of $IrMe_3(CO)(TMED)$ [37].

β-Diketonate Derivatives of Ir(III)

All the known methods of the synthesis of Ir(III) β-diketonates involve mainly chloride or hydroxo complexes of Ir(III) as the starting compounds. The ions Cl^-- and OH^- compete with the chelate-forming ligand for the coordination site. In addition, these ions in iridium compounds are inert in substitution processes, which leads to the formation of various polymeric and polynuclear forms in the reaction mixture.

For instance, the yield of $Ir(acac)_3$ from the interaction of K_3IrCl_6 with potassium acetylacetonate in the aqueous solution [38] is only 10%. The isolated complexes having the composition $K_2Ir(acac)_2XY.nH_2O$ (where X, Y - Cl^-, OH^-) contain only oxygen-bound acetylacetonate groups The yield 20% of $Ir(acac)_3$ was achieved in the interaction of aquated iridium trichloride with a 10-fold excess of the ligand [39]. The authors of [39] succeeded in isolating the complexes having the composition $Ir(acac)_3(p\text{-}CH_3C_6H_4NH_2)$, $Ir_2(acac)_6$, $Ir(acac)_3(NH_3)$, $Ir(acac)(Pr^nNH_2)$; they demonstrated that the reaction of $IrCl_3.nH_2O$ with acetylacetone leads mainly to the formation of the complexes containing γ-C-bound acetylacetonate groups, which is the reason of the low yield of $Ir(acac)_3$. By present, the yield of $Ir(acac)_3$ not higher than 50% has been achieved using the chlorocomplexes of Ir(III) in reaction with acetylacetone [40]. Ir(III) pivaloylacetonate was obtained with a small yield using a similar method [41].

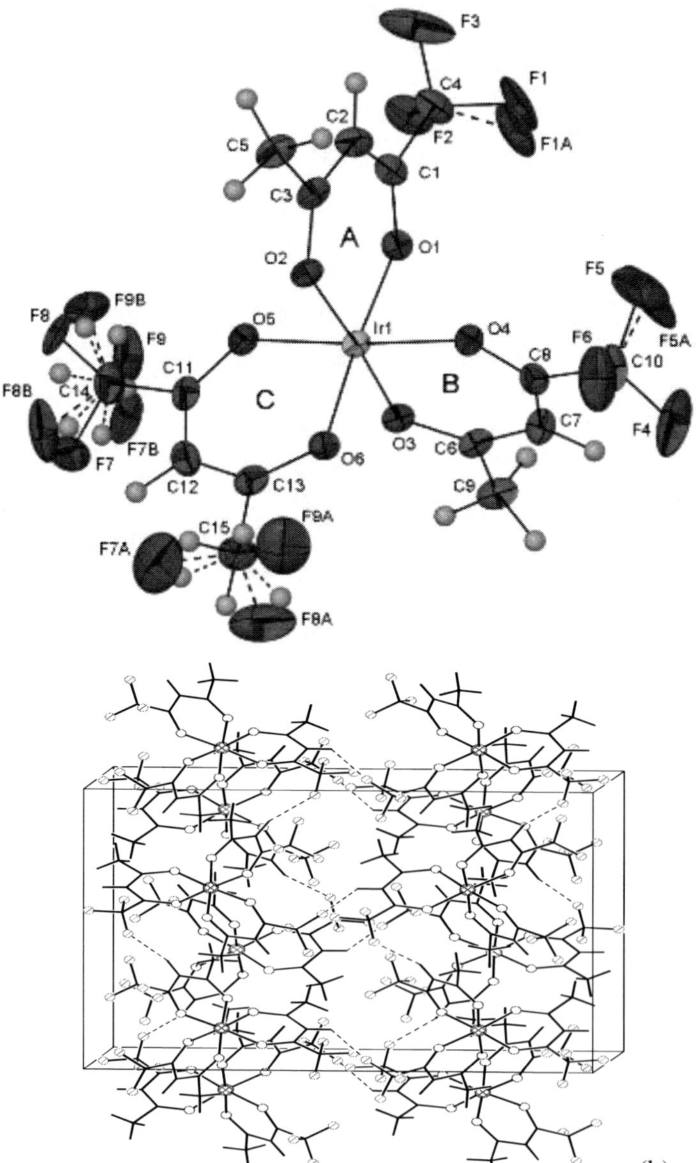

(a)

(b)

Figure 1. Structure of *trans*-$Ir(tfac)_3$ molecule, with atom numbering (*a*), and molecular packing in the crystal along the direction (100) (*b*).

The dissolution of freshly precipitated Ir(III) hydroxide in perchloric acid and the interaction of hexaaqua anion Ir(III) $[Ir(H_2O)_6]^{3+}$ with β-diketone lead to the low yield of the final product because polynuclear forms appear [42].

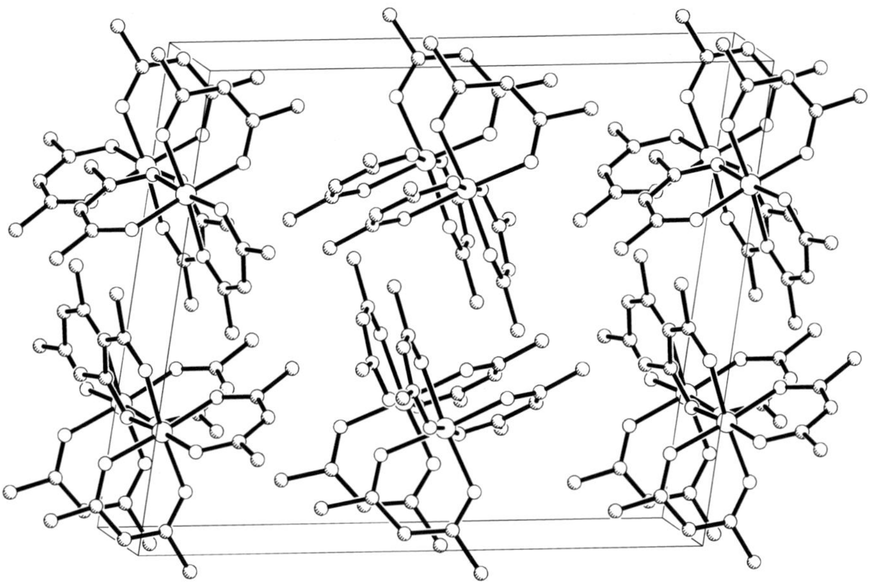

(a)

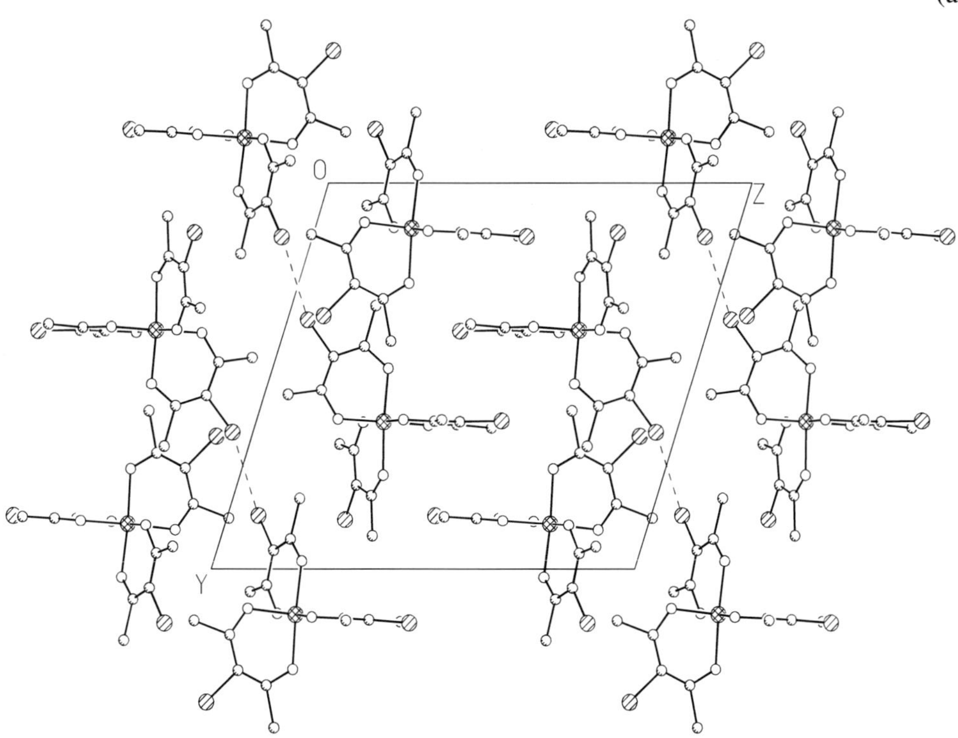

(b)

Figure 2. (Continued)

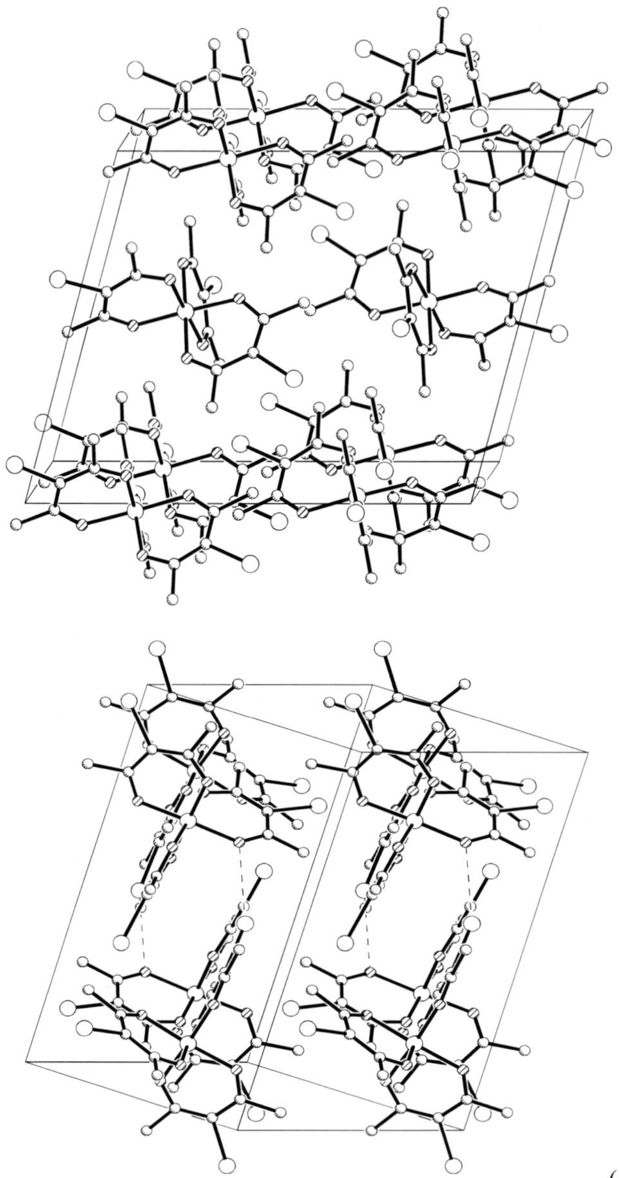

(c)

(d)

Figure 2. Packing of molecules in the crystals of the complexes $Ir(acac)_3$ (*a*), $Ir(acacCl)_3$ (*b*), $Ir(acacBr)_3$ (*c*), $Ir(acacI)_3$ (*d*).

In our investigations, we developed the method to synthesize Ir(III) β-diketonates from iridium hexafluorocomplexes with subsequent aquation of the $[IrF_6]^{3+}$ form. When implementing the synthesis procedure, a known fact was used: the lability of fluoride ion in the case of the low degrees of metaloxidation [43]. Taking into account also its low competitive ability in comparison with aqua ions, the proposed synthesis method gives almost quantitative yield of final products. The yield of $Ir(acac)_3$ is 90-95% [43]. The same procedure was used to obtain other Ir(III) β-diketonates with non-fluorinated substituents in the ligand [43-45].

The synthesis of Ir(III) complexes with fluorine-containing β-diketonates was carried out according to the procedure similar to that described above for $Ir(acac)_3$, but longer time was necessary (1-2 days, under heating with the backflow condenser). As a rule, zone sublimation in the vacuum gradient furnace was sued as the purification method for Ir(III) β-diketonates; the yields of compounds were 70-90% [43-45]. For the compound $Ir(tfac)_3$, separation of the structural isomers was carried out by means of column chromatography on aluminium oxide; a mixture of heptane with benzene was used as the solvent. According to the ^1H NMR spectroscopic data, the product after synthesis is a mixture of the *cis*-form (20%) and *trans*-form (80%) [46].

The purity of the isolated iridium chelates was confirmed by means of elemental analysis for C, H, F; IR, NMR spectroscopy and mass spectrometry, and on the basis of measured melting points [47, 48]. All the synthesized compounds are yellow or yellowish-orange solid crystal substances; they are stable against external action. These compounds get crystallized in the lattices of molecular type. In these crystals, the molecules are bound by Van der Waals interactions, that is, the energy of intermolecular bonds is essentially lower than the energy of intramolecular bonds. This circumstance provides one of the unique properties of molecular crystals, namely their volatility, that is, the ability to pass into the gas phase without decomposition. In turn, volatility depends mainly on the character of molecular packing in the crystal, which is determined directly by the structure of molecules. According to the single crystal X-ray diffraction analysis, the structures of the compounds are composed of neutral molecules; iridium atom forms the common octahedral coordination sphere with three beta-diketones as ligands (Figure 1 *a*, *b*) [46].

According to our data, the compounds are stored at room temperature for several years without any changes in their physicochemical parameters. Fluorinated complexes are less stable during storage.

The authors were also the first to obtain iridium compounds containing the substituents in the γC-position of acetylacetonate ligand, like $Ir(CH_3\text{-}CO\text{-}CX\text{-}CO\text{-}CH_3)_3$ (X – F, Cl, Br, I) according to reaction [49]:

$$Ir(acac)_{3(s)} + nHX + nH_2O \longrightarrow Ir(acacX)n(acac)_{3\text{-}n(s)} + 2nH_2O$$

with almost quantitative transformation of $Ir(acac)_3$ into γC-halogen-substituted acetylacetonates, though the quantitative formation of $Ir(acacX)_3$ seems to be impossible in this case because mono- and disubstituted Ir(III) acetylacetonates are formed along with triple-substituted ones. The complexes were isolated by means of chromatography with the column filled with Al_2O_3. The yields of compounds were 10 to 40%. All the synthesized compounds are yellow crystalline substances.

The structures of γC-halogen-substituted iridium acetylacetonates belong to the molecular type, too (Figure 2) [49]. In all the complexes, the central Ir atom has slightly distorted octahedral coordination by six oxygen atoms of three β-diketonate ligands. Each molecule in the structures is surrounded by the ten nearest neighboring ones, so that the distances Ir…Ir are estimated for $Ir(acac)_3$ to be > 7,49 Å; for $Ir(acacCl)_3$ - 7,98-10,08 Å; for $Ir(acacBr)_3$ and $Ir(acacI)_3$ - > 8,10 Å. For compound $Ir(acacCl)_3$, the shortest intermolecular contact Cl(2)…Cl(4), is to be stressed; it is 3,26 Å long. Due to this interaction, crystallographically independent molecules are bound in pseudo-dimers. In general, the introduction of halogen into the γC-position does not have any essential effect on the crystal chemical characteristics.

In [50], the methods of Ir(cod)(L) synthesis are described, for L = acac, tfac, hfac. The reactions are carried out through the interaction of [Ir(OCH_3)(cod)] and β-diketone in pentane solution at low temperature in the atmosphere of nitrogen according to the scheme:

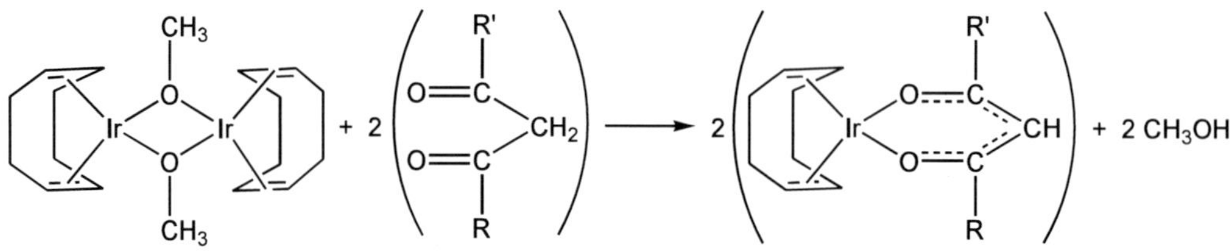

The complexes are purified by sublimation or recrystallization from pentane solution.

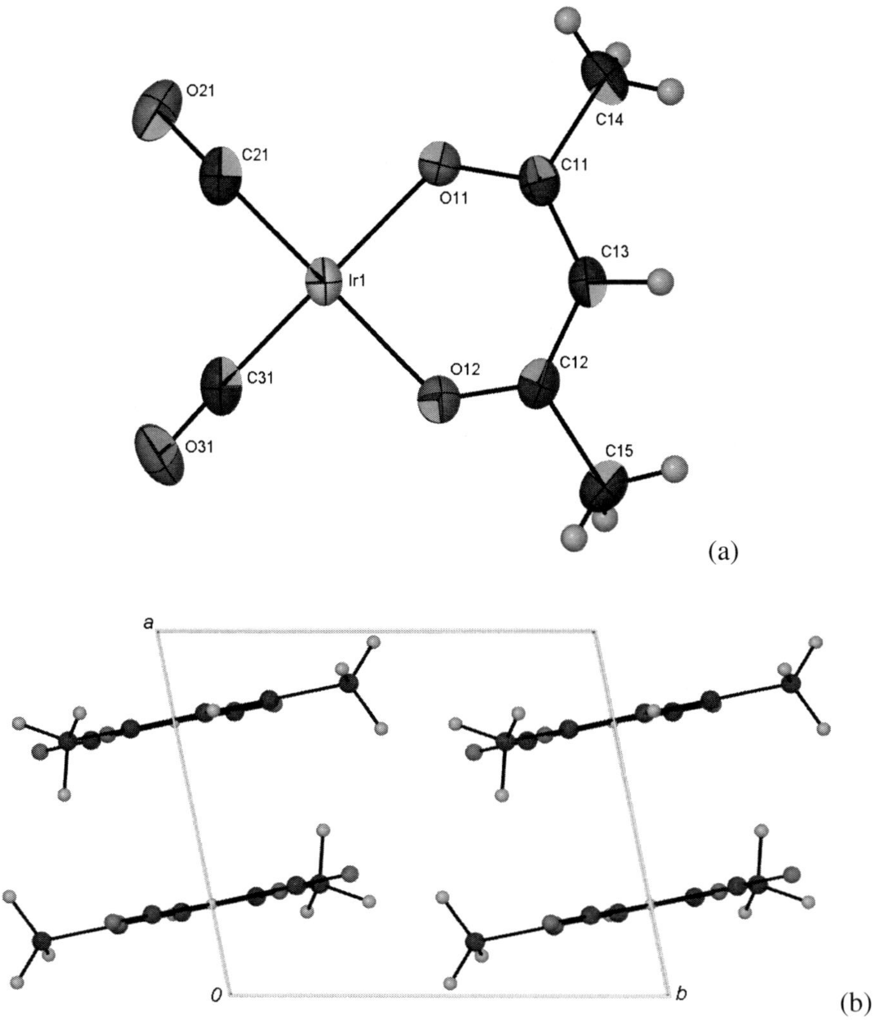

Figure 3. The structure of the molecule Ir$(CO)_2$(acac) (*a*) and the view of the crystal structure along direction *c* (*b*).

Cyclopentadienyl Derivatives of Iridium

By boiling iridium carbonylhalides with cyclopentadiene in excess in organic solvents, $Ir(CO)_2(Cp)$ is obtained [51]. A similar complex $Ir(CO)_2(Cp')$ can be obtained by passing CO through the solution of $[IrCl_2Cp']_2$ in dichloromethane and adding zinc powder [52]. After boiling $Ir(CO)_2(Cp')$ in butanol for one week at 170°C, the dimeric product $[Ir(CO)_2(Cp')]_2$ is formed. The carbonyl groups in it play the part of bridging ligands. The four-membered cycle Ir-C(1)-Ir'-C(1)' has a flat structure [52]. Complexes $[IrCl_2Cp]_2$ and $[IrCl_2Cp^*]_2$ can be obtained in the interaction of hexachloroiridium acid and cyclopentadienyl ligands under heating in methanol solution in the inert atmosphere. Further cooling on the reaction mixture leads to the deposition of crystalline product [53]. The complexes were identified according to the data of IR, NMR spectroscopy. In the interaction of $[Ir(cod)Cl_2]_2$ and NaCp' in tetrahydrofuran solution in the inert atmosphere under heating, the complex Ir(cod)(Cp') is formed. The compound sublimates at a temperature of 50°C and a pressure of 0,01 torr [54]. The spectral characteristics of Ir(cod)(Cp') were also presented in [54].

For a number of Ir(I, III) compounds, we modified the known synthesis methods [52-54] and developed the procedures of the synthesis using hexachloroiridium acid as the starting reagent [55] according to the schemes:

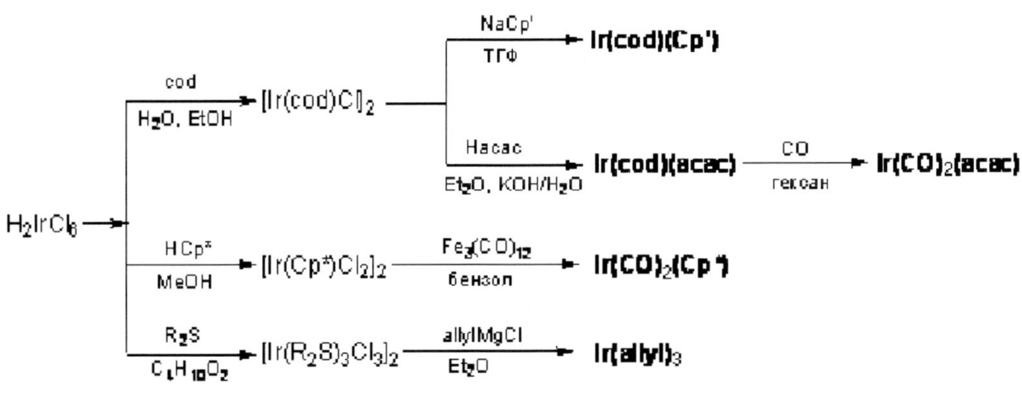

The synthesized compounds are crystal substances of white (Ir(cod)(Cp'), $Ir(allyl)_3$), yellow (Ir(cod)(acac), $Ir(CO)_2(Cp^*)$) or brown ($Ir(CO)_2(acac)$) colour. They are well soluble in organic solvents and insoluble in water. After purification, they are stable during storage for a long time. The mass spectra of these complexes contain intense peaks of molecular ions, that is, all the compounds pass into the gas phase in the form of monomers within the studied temperature ranges (as a rule, 60-120°C). No peaks with masses larger than the mass of the molecular peak were detected. The relative intensities of ion peaks in the mass spectra do not change which points to thermal stability of the compounds under the experimental conditions.

The structures of $Ir(CO)_2(acac)$ and Ir(cod)(Cp') are composed of neutral isolated mononuclear molecules. In the crystal of $Ir(CO)_2(acac)$, molecules are packed in piles so that the planes of coordination squares turn out to be parallel, and the distances between the nearest neighbors in a pile are Ir…Ir 3,242 and 3,260 Å. The closest neighboring molecules in a pile are packed with turning by 180° (Figure 3) [55].

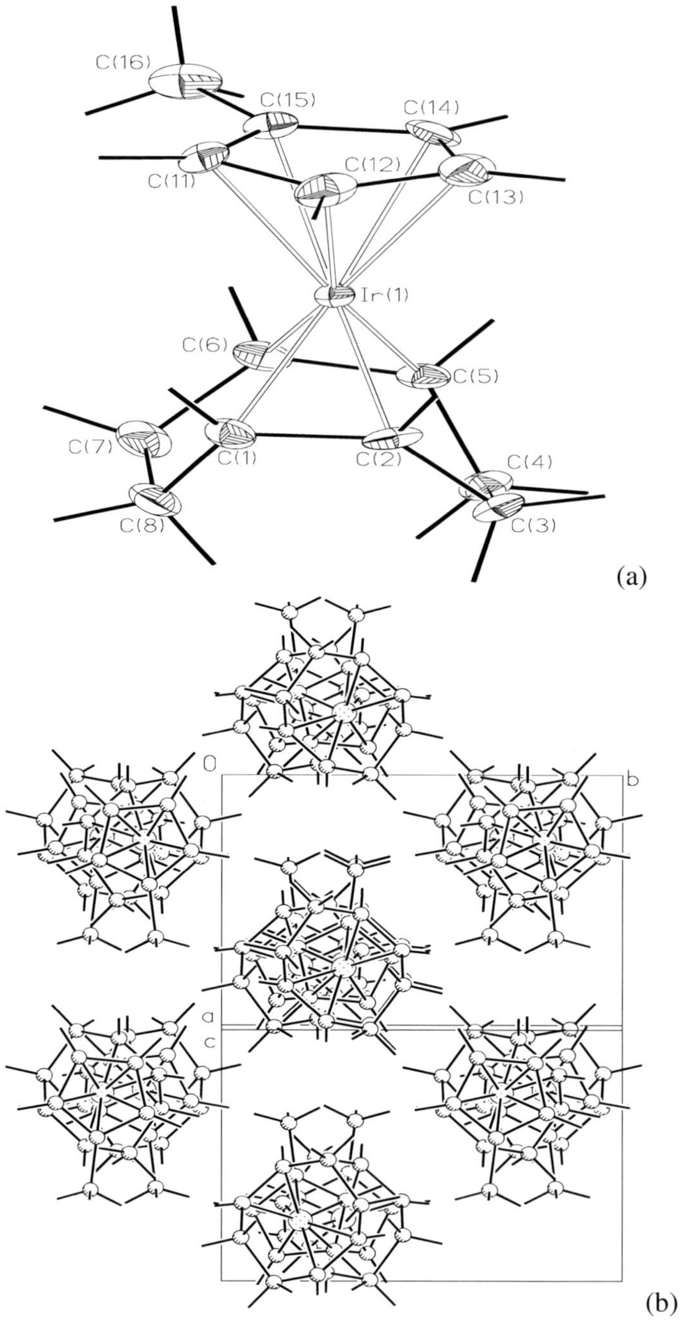

Figure 4. **Structure of Ir(cod)(Cp') molecule** (*a*) and the projection of the crystal lattice on the plane (-101) (*b*).

The molecule of Ir(cod)(Cp') does not possess any elements of symmetry (Figure 4). The metal atom coordinates carbon atoms of two cyclic ligands – 5-methylcyclopentadienyl ion (Cp') and 1,5-cyclooctadiene (cod). Only Van der Waals interactions bind the molecules in the crystal [55].

Allyl and Amidinate Derivatives of Iridium

The synthesis of *tris*-(η-allyl)Ir(III) was described only in few works; the yield of the final product does not exceed 20% [56]. $Ir(allyl)_3$ is obtained either through the interaction of magnesium allyl chloride with Ir(III) acetylacetonate (yield: 20%) or with the chlorinated complex $IrCl_3.xH_2O$ (yield: 10%) [56]. The white crystals of the compound easily sublimate at room temperature and a pressure of 15 torr. The compound decomposes in the air.

The interaction of $[IrH_3(PPh_3)_3]$, $[IrHCl_2(PPh_3)_3]$–NEt and $[IrHCl_2(PPh_3)_3]$ with *N,N'*-diphenylamide compounds PhN=C(R)–NHPh (R = H, Me, Et or Ph) in boiling benzene or toluene results in the formation of the crystalline amidinate derivatives of iridium $[IrH_2\{PhNC(R)NPh\}(PPh_3)_2]$, $[IrH(Cl)\{PhNC(R)NPh\}(PPh_3)_2]$ and $[IrCl_2\{PhNC(Ph)NPh\}(PPh_3)_2]$, respectively [57]. The compounds were characterized by means of elemental analysis, IR, NMR spectroscopy, and melting points.

Complexes of Iridium with (S,S)-ligands

A substantial number of different types of sulfur-containing ligands forming volatile chelates with many metals are known. These first of all include dithiocarbamates having the group R_2NCS_2; dithiophosphates, dithiophosphinates, thio-β-diketonates and dithio-β-diketonates.

By the present moment, 12 different Ir(III) dithiocarbamates ($Ir(R_2dtc)_3$) have been synthesized. For example, Ir(III) diethyldithiocarbamate was synthesized using metal chloride according to reaction [58]:

$$MCl_n.xH_2O + n(Na(EtNCS_2).3H_2O) \rightarrow M(EtNCS_2)_n + nNaCl + (3n+x)H_2O.$$

All the compounds are solid substances with melting points 98-300°C. The complexes are well soluble in organic solvents, insoluble in water, stable in the air, sublimate in vacuum. According to the single crystal X-ray diffraction analysis, the crystals of $Ir(Et_2dtc)_3$ belong to the space group *C2/c* with monoclinic type; the average Ir–S distance is 2.36 Å, the manner in which the molecules are packed in the crystal is shown in Figure 5 [59].

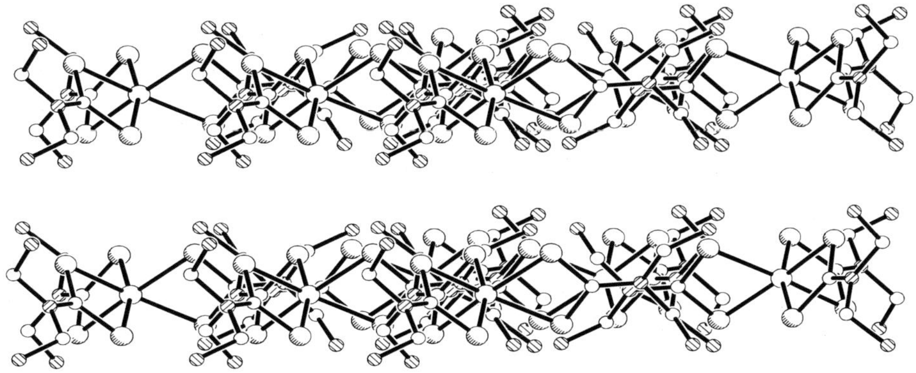

Figure 5. Packing of $Ir(Et_2dtc)_3$ molecules in the crystal.

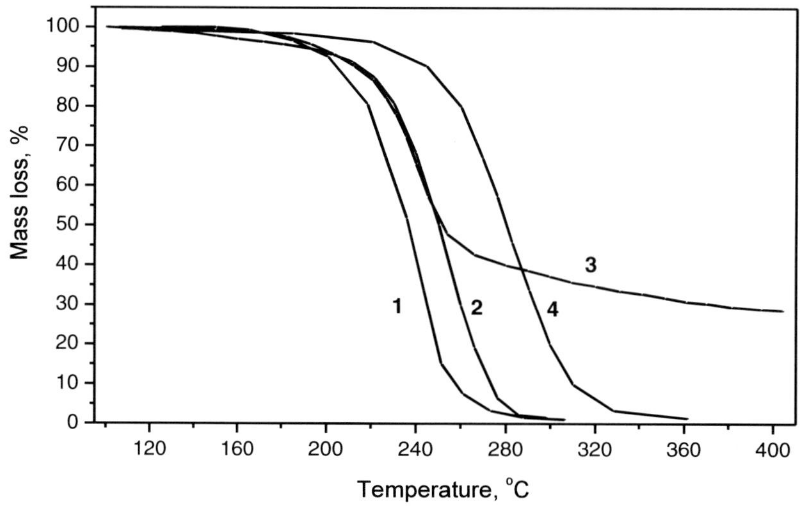

Figure 6. TG curves for Ir(III) acetylacetonate and halogen-substituted Ir(III) acetylacetonates in helium flow (open crucible): 1 – $Ir(acacI)_3$, 2 – $Ir(acacCl)_3$, 3 – $Ir(acacBr)_3$, 4 – $Ir(acac)_3$.

Table 1. Results of thermal investigations of Ir(III) β-diketonates

Compound	T_{decomp}*, °C (ΔT = ±5) (condensed phase)		
	DTA/He	DTA/air	DTA/O_2
$Ir(acac)_3$	350	310	252
$Ir(tfac)_3$	300	288	208
$Ir(ptac)_3$	-	312	202

* $T_{decomp.}$ is temperature of decomposition onset.

Table 2. Result of thermal analysis of $Ir(acac)_3$ and halogen-substituted Ir(III) acetylacetonates

Compound	$T_{decomp.}$, °C (DTA, close crucible)	T (50%)*, °C (TG, open crucible)	$T_{melt.}$, °C
$Ir(acac)_3$	350	310	269
$Ir(acacCl)_3$	290	250	278
$Ir(acacBr)_3$	200	-	200 decomp.
$Ir(acacI)_3$	270	245	258

* Temperature corresponding to the 50 % mass loss.

Thermal Properties of Volatile Iridium Complexes

Thermal Stability of the Compounds in the Condensed State

As a rule, during the deposition of films or coatings, precursors are kept for a long time in the condensed phase at a high temperature. Modern technologies require the stability of precursors at the sublimation temperature for several days. In this case, very often the limiting parameter defining the possibility to use a compound in MOCVD processes is its thermal stability in the condensed state, that is, the stability under heating to a definite temperature without any changes of the chemical (component) composition of the compound in the solid phase. As a rule, thermal stability is characterized by the temperature, recorded by means of thermal analysis, at which the change of the component composition in the condensed phase starts. The most widely used methods for the determination of thermal stability of compounds in the condensed state are the methods of complex thermal analysis (DTA, DSC, TG, DTG). The theory and methodology of these methods have been described in detail in monograph [60]. The application of these methods brings about some difficulties connected with the occurrence of two simultaneous processes: mass loss due to decomposition and sublimation (evaporation) during heating, which complicates the interpretation of results. A compound often gets sublimated (evaporated) from the crucible even before the decomposition process starts to be recorded, and in this case crucibles with tight caps may be used, or a thermal diluent which is usually Al_2O_3 at the mass ratio to the weighed portion of the complex compound 10:1. However, for highly volatile compounds even these precautions may be insufficient. A specific feature of the methods of complex thermal analysis is the possibility to determine the temperature of the onset of the interaction between the complex compound and the reactant gases (for example, hydrogen or oxygen). The data obtained by us using DTA and TG procedures for Ir(III) β-diketonates are presented in [47]. Using the tightly closed crucibles, temperatures of the onset of decomposition were measured for a number of Ir(III) β-diketonates. The results of thermal studies are shown in Table 1. It was concluded from the TG curves that the complexes with fluorinated substituents in the ligand exhibit higher sublimation ability than the alkyl-substituted Ir(III) β-diketonates.

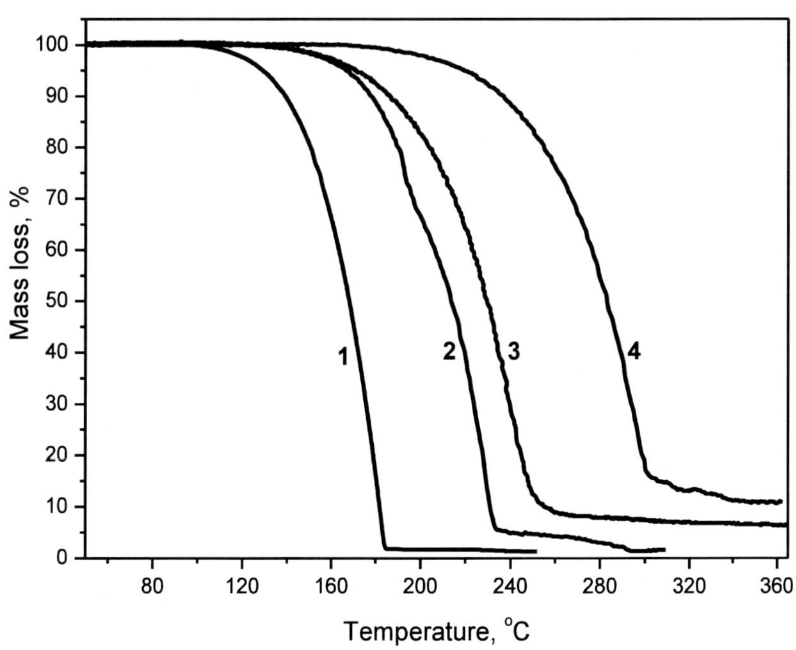

Figure 7. TG curves in helium: 1 – Ir(cod)(Cp'), 2 – $Ir(CO)_2$(acac), 3 – $r(CO)_2$(Cp*), 4 – Ir(cod)(acac).

On the basis of TG curves recorded for halogen-substituted iridium(III) acetylacetonates [49] in helium flow, it may be concluded that their volatility increases in comparison with the initial Ir(III) *tris*-acetylacetonate: the TG curve of the latter lies in the region of higher temperatures (Figure 6). As far as thermal stability is concerned, it decreases in comparison with $Ir(acac)_3$ for the entire set of the studied γ-halogen-substituted It(III) acetylacetonates (Table 2).

For Ir(III) diethyldithiocarbamate, almost 100 % mass loss in observed during heating in vacuum up to 340°C (residue: 10%), in helium flow the complex is transformed into the gas phase with decomposition [61]. According to the DTA data, $T_{melt.}$ = 265°C, ($T_{melt.}$ (Kefler table) = 265°C), $T_{decomp.}$ (He) = 320°C, T (50% mass loss, vacuum) = 270°C.

The heteroligand complexes of Ir(I) when heated in helium atmosphere during recording pass into the gas phase almost without decomposition within temperature range 160-300°C. The most volatile compound in the presented row is Ir(cod)(Cp') (Figure 7 [55]).

For the 50% mass loss of complex compounds under identical experimental conditions, the following qualitative volatility sequence was established:

Ir(cod)(Cp') (168°C) > $Ir(CO)_2$(acac) (213°C) > $Ir(CO)_2$(Cp*) (229°C) > Ir(cod)(acac) (283°C).

For Ir(cod)(Cp') and Ir(cod)(acac), the DTA curve exhibits endothermic peaks corresponding to melting processes: at temperatures 40°C and 155°C [55].

Temperature Dependencies of Saturated Vapour Pressure for Volatile Iridium Complexes with Organic Ligands

The phenomenon of volatility, that is, the ability of compounds to pass into the vapour phase without decomposition, was discovered in the end of the 19th century. The quantitative characteristic of volatility is vapour pressure at a fixed temperature. The existing methods to determine saturated vapour pressure are rather diverse: static, quasi-static, dynamic (flow method and Langmuir method), kinetic methods (various modifications of Knudsen effusion method). The principles of the methods were described in detail in [62]. Experimentally measured temperature dependencies of saturated vapour pressure are presented in the form of equations $\ln P = B - A/T$, where A and B coefficients are linked with the thermodynamic parameters of vaporization processes in the following manner: $B = \Delta S^{o}_{T*}/R$, $A = \Delta H_{T*}/R$. Thermodynamic parameters enthalpy ΔH_{T*} and entropy ΔS^{o}_{T*} relate in the case under consideration to the intermediate temperature range (T* is the middle of the temperature range) within which vapour pressure is measured. The volatility of complexes is observed in the case if the energy of intermolecular bonds (Van der Waals bonds, electrostatic and dipole interaction between monomeric molecules in the crystal) is much lower than the energy of intramolecular bonds (bonds inside a molecule). So, volatility and thermal stability of a compound are first of all connected with its structural characteristics and electronic structure.

We established the major regularities of vapour pressure changes for the most well studied class of volatile compounds – beta-diketonate derivatives of metals, as a function of the structure of the β-diketonate ligand involved [47]: this is the nature of the central atom, terminal substituents and donor atoms; steric and inductive effects affecting the external electron surroundings of complex molecules and some other factors are to be taken into account as well. However, depending on the methods used to measure vapour pressure, results obtained for one and the same compound may differ from each other substantially. A method is chosen taking into account the specificity of a compound under investigation.

Static Methods

In the static methods, measurement of vapour pressure is carried out in a closed volume; temperature is maintained constant, so that the equilibrium between the condensed phase and the vapour was established. The equilibrium pressure is determined either directly with the help of a measuring tool, which is a mercury, ionization, spiral, membrane or another kind manometer (direct methods) or by determining the amount of the substance in a definite volume (indirect methods).

Flow Method (Dynamic)

The essence of the method is as follows. A flow of inert gas is passed over the heated substance to carry the vapour of the substance into condenser. The amount of the transferred substance is determined with the help of gravimetry, chemical analysis or radiation method. The desired result may be obtained for different inert gas flow rates if one knows total pressure $P_{total.}$, the amount of transferred substance n and inert gas N.

Knudsen Method (Kinetic)

The method is based on measuring the rate of vapour flow through a small orifice from the space containing saturated vapour. Knudsen method has a variety of versions due to its specificity; it is widely used to measure low vapour pressure values. The method is very sensitive to the type of substance, its aggregate state, vapour composition, reactivity.

Vapour Pressure of Volatile Iridium Complexes

In [63], differential scanning calorimetry was used to determine the enthalpies of the reaction of solid complexes (acac)Ir(olefine)$_2$ (olefine = ethylene (Et), propylene (Pr), vinyl chloride (VCl), vinyl acetate (VAc) etc.) with $CO_{(gas)}$ according to reaction:

$$\text{(acac)Ir(olefine)}_2\text{ (s)} + 2\text{CO (gas)} \longrightarrow \text{(acac)Ir(CO)}_2\text{ (s)} + 2\text{olefine (gas)}$$

Table 3. Enthalpies of sublimation for Ir(L)$_2$(acac) (lg(P, torr) = B–A/T)

Compound	B	A	ΔH_{T*}, kJ/mol	ΔT, °C
$Ir(CO)_2(acac)$	12,2	4821±68	92,0±1,3	13–52
$Ir(Et)_2(acac)$	10,6	4324±238	82,8±4,2	10–28
$Ir(Pr)_2(acac)$	12,8	4697±71	90,0±1,3	-4–25
$Ir(VCl)(acac)_2$	11,3	4684±225	89,5±4,6	8–25
$Ir(VAc)_2(acac)$	14,1	6284±163	120,5±3,3	52–70

Enthalpies of sublimation were calculated from the obtained values. The results for some heteroligand complexes of iridium are presented in Table 3.

We studied the vapour pressure of the β-diketonate derivatives of Ir(III). Using the static method with membrane gauge manometer we measured vapour pressure for a number of Ir(III) β-diketonates in the region of relatively high pressures [47, 48]. Figure 8 shows the dependencies of vapour pressure on reciprocal temperature for the studied complexes. Thermodynamic parameters of vaporization processes are shown in Table 4. The temperature dependencies of saturated vapour pressure for iridium complexes depending on the type of substituent in the ligand display the trends similar to other β-diketonate derivatives of metals: the introduction of fluorinated substituents into the chelate increases the volatility of the complex. The authors of [47, 48] paid attention to the fact that the absolute values of ΔH_{T*} and ΔS_{T*}^{o} characterizing sublimation or evaporation decrease with the introduction of CF_3-group into the ligand. In this situation, a decrease in ΔH_{T*} that should cause a sharp rise of the saturated vapour pressure is compensated by the corresponding decrease in ΔS_{T*}^{o}. The compensation dependencies of this kind are observed separately for each type of ligand; the degree of metal oxidation does not play any role [47]. During melting (Table 4), substantial convergence of ΔH_{T*} and especially ΔS_{T*}^{o} occurs. The range of vapour pressure measurement in evaporation processes is much narrower [47].

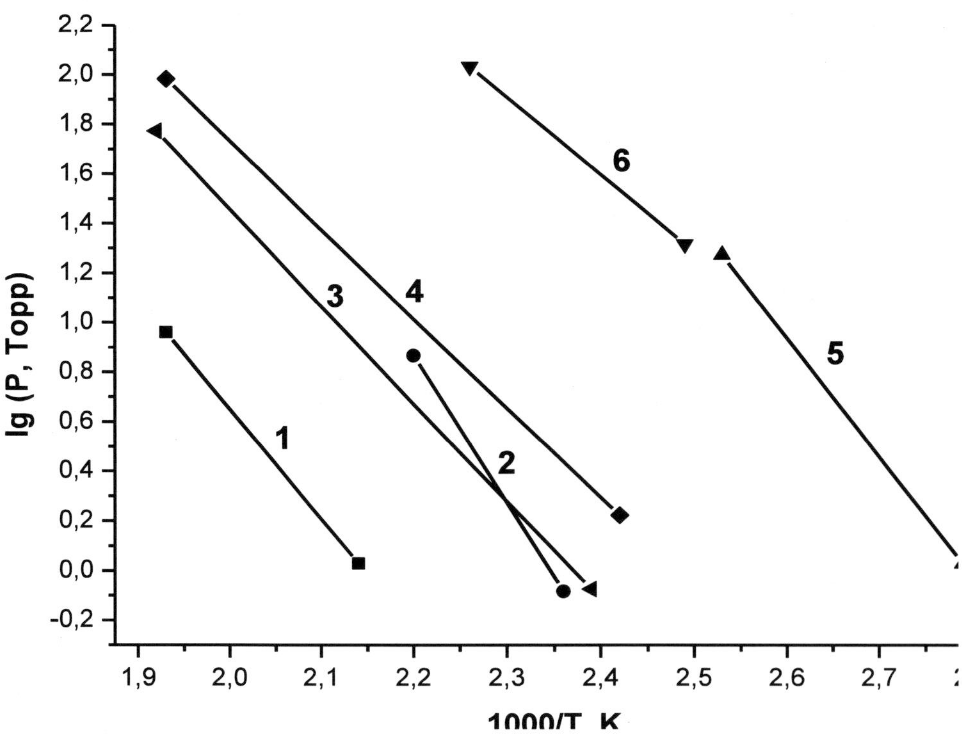

Figure 8. Temperatuer dependencies of saturated vapour pressure for Ir(III) β-diketonates: 1 – $Ir(acac)_3$, 2 – $Ir(tfac)_3$, 3 – $Ir(thd)_3$, 4 – $Ir(ptac)_3$, 5 – $Ir(hfac)_3$ (sub.), 6 – $Ir(hfac)_3$ (evap.).

Table 4. Thermodynamic parameters of vaporization processes for Ir(III) β-diketonates

Compound	Process	ΔH_{T*}, kJ/mol	ΔS_{T*}^{o}, J/mol·K
$Ir(acac)_3$	sub.	86,6±2,1	130,1±4,2
$Ir(tfac)_3$	sub.	116,3±8,7	218,8±17,6
$Ir(hfac)_3$	sub.	88,7±5,8	193,7±9,6
$Ir(hfac)_3$	evap.	56,5±0,8	99,2±2,1
$Ir(ptac)_3$	evap.	68,6±0,8	115,5±1,7
$Ir(thd)_3$	evap.	74,1±2,1	115,9±3,8

On the basis of the mentioned data, the sequence of volatility changes for iridium(III) complexes depending on the type of substituent in the ligand was determined:

$Ir(acac)_3 < Ir(thd)_3$, $Ir(tfac)_3 < Ir(ptac)_3 < Ir(hfac)_3$.

For Ir(III) diethyldithiocarbamate $Ir(dtc)_3$, the flow method was used to obtain the equation for saturated vapour pressure: $\ln(P, \text{atm.}) = -21063/(T, \text{K}) + 29{,}7$ and thermodynamic parameters of sublimation process were calculated: $\Delta H_{T*} = 175{,}1 \pm 4{,}0$ (kJ/mol); $\Delta S°_{T*} = 245{,}8 \pm 7{,}8$ J/mol·K.

The *P*/*T* dependencies for a number of heteroligand complexes of Ir(I) are presented in Figure 9. Thermodynamic parameters of vaporization processes are shown in Table 5 [55].

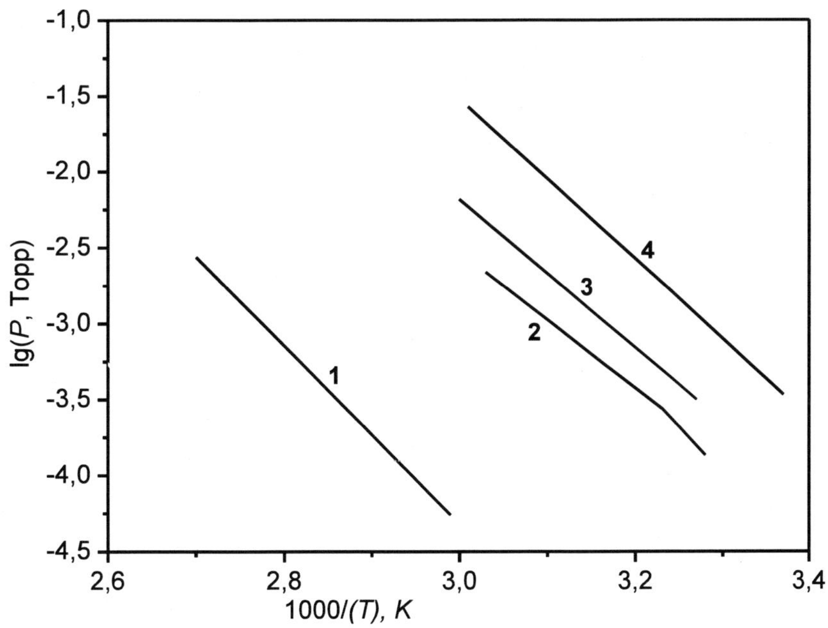

Figure 9. *P*/*T* dependencies for iridium(I) complexes: 1 – Ir(cod)(acac), 2 – Ir(cod)(Cp'), 3 – $Ir(CO)_2(acac)$, 4 – $Ir(CO)_2(Cp^*)$.

Table 5. Thermodynamic parameters of vaporization processes for heteroligand Ir(I) complexes

Compound	Process	ΔT, °C	ln(*P*, atm) = –A/(*T*, K)+B		ΔH_{T*}, kJ/mol	$\Delta S°_{T*}$, J/mol·K
			A	B		
Ir(cod)(Cp')	**sub.**	**31–38**	**33,4**	**14973**	124,6±5,0	279±16
	evap.	**38–77**	**19,4**	**10580**	88,1±1,3	161,2±4,2
$Ir(CO)_2(Cp^*)$	**sub.**	**24–59**	**27,9**	**12627**	105,0±3,4	232±11
Ir(cod)(acac)	**sub.**	**62–97**	**23,7**	**13432**	111,7±1,7	197,5±5,0
$Ir(CO)_2$(acac)	**sub.**	**33–60**	22,4	11293	94,1±2,7	186,1±7,8

Among the heteroligand Ir(I) compounds the most volatile complex within the studied temperature ranges is $Ir(CO)_2(Cp^*)$, while the least volatile one is Ir(cod)(acac).

Thermal Stability of the Vapour of Volatile Iridium Complexes

The term “thermal stability of vapour” (thermolysis of vapour) means the threshold temperature at which the heterophase processes start through which the vapour of initial compound is transformed into the products of thermal decomposition (gaseous and condensed) on the heated surface. Unlike for the condensed phase, in this case the decomposition of individual molecules adsorbed on the surface is investigated.

A unique procedure described in [64] deals with the studies of thermal transformations of the compounds. It allows one to obtain the data on the processes of thermal decomposition of vapour in the controlled atmosphere (vacuum, oxygen, hydrogen and other reactive additives). An original system has been developed on the basis of the CVD reactor connected with the mass spectrometer; the latter instrument is equipped with the molecular beam source. The system allows *in situ* identification of the gas phase composition. This set-up is a version of the two-chamber Knudsen cell. The first chamber into which the initial compound is put acts as an evaporator. The second chamber connected with the heated pipeline with the evaporator is a thermal reactor. The molecular beam passes from the reactor through the effusion orifice directly into the ionization chamber of the mass spectrometer, which serves as the analyzer of the gas phase composition. Additionally, the reaction gas can be supplied to the input of the thermal reactor, which provides the possibility to study its effect on the processes of vapour decomposition of compounds under investigation. The flow regime is molecular in all the parts of the system, which excludes the possibility of the reactions of initial compounds with the products of decomposition, and the effect of diffusion on the kinetics of the reaction.

Table 6. Investigation of thermal destruction of Ir(III) β-diketonates vapour

Compound	$T_{decomp.}$, °C (ΔT = ±10) (gaseous phase)		
	vacuum	O_2	H_2
$Ir(acac)_3$	405	225	410
$Ir(hfac)_3$	430	-	-

The procedure in use allows one to investigate the compounds within a broad temperature range and to record both the primary reaction products and the profound degrees of destruction observing the qualitative changes in the composition of the gaseous products of thermolysis. The information about the process is recorded in the form of the curves showing changes of the intensities of ion currents in the mass spectrum during the stepwise rise of reactor temperature. The mass spectrometer records these changes in the real time mode. Below the specific results of the studies of vapour thermolysis of volatile iridium complexes will be considered.

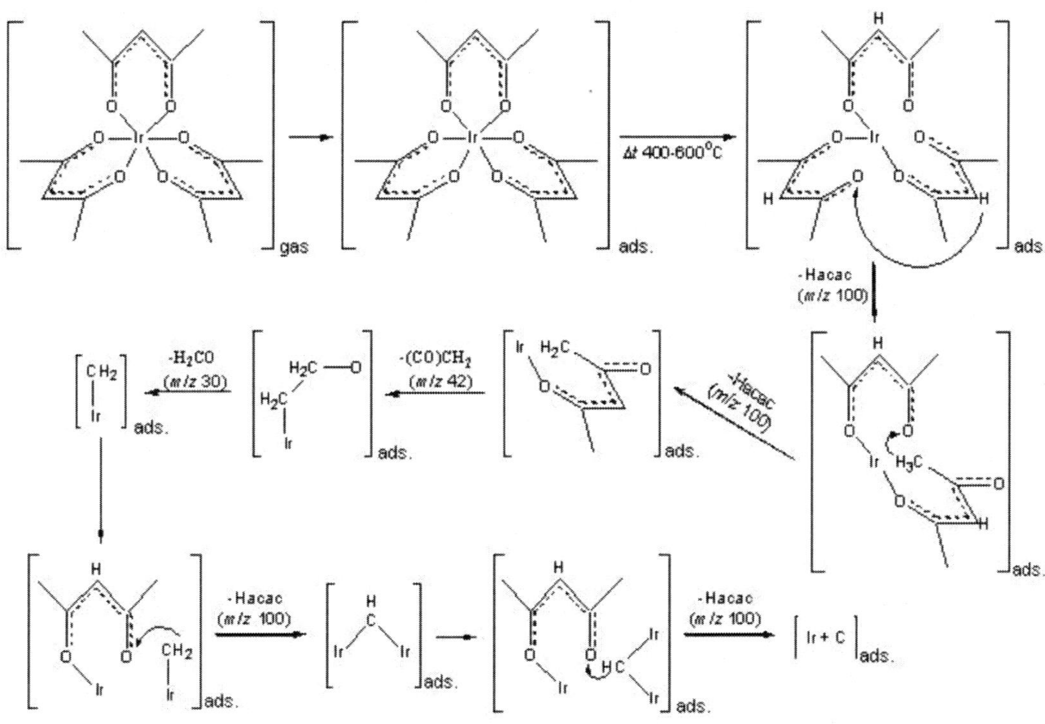

Figure 10. Scheme of possible routes of chemical transformations of $Ir(acac)_3$ vapour on heated surface in vacuum.

For $Ir(acac)_3$, thermal stability of the vapour was studied in vacuum, oxygen and hydrogen; for $Ir(hfac)_3$ – in vacuum. Temperature intensities of ion currents corresponding to the major gaseous products of decomposition were recorded, and the threshold temperatures of thermal decomposition were determined [65]. The data are presented in Table 6.

Relying on the analysis of temperature dependencies of the gas phase composition [65], we may represent thermal decomposition of Ir(III) acetylacetonate vapour in vacuum (10^{-4}

torr) within temperature range 200-600°C by the scheme shown in Figure 10 [66]. At the first stage at 405°C acetylacetone is released into the gas phase, then within temperature range up to 600°C a number of carbon-containing fragments are recorded in the gas phase: CH_2CO, H_2CO etc. After the introduction of hydrogen, the stability of vapour slightly increases (410°C), but the scheme of decomposition does not change: acetylacetone molecules and the same organic species but in smaller amounts are observed in reaction products.

The mechanism of thermolysis of $Ir(acac)_3$ vapour changes sharply after the addition of oxygen into the system: the threshold temperature decreases to 220°C, and the major reaction products detected in the gas phase are CO_2 and H_2O, that is, rapid interaction of the chelate vapour with oxygen occurs, so that free metal is formed on reactor walls, and the products of profound oxidation of ligand appear (Figure 11). In this case, the reaction rate is affected to a higher extent by oxygen concentration in the reaction zone rather than by temperature.

The temperature dependencies of the gas phase composition during thermal decomposition of $Ir(CO)_2(acac)$ vapour in vacuum (10^{-4} torr) are presented in Figure 12 [67]. Analysis of mass spectral data showed that the major reaction products are acetylacetone, CO, CO_2 and C_2H_3O, which is presumably a radical species. The formation of acetylacetone is possible for this complex only in the case of the interaction between at least two molecules, that is, thermolysis in vacuum occurs through the bimolecular mechanism with the detachment of carbonyl and opening of the chelate ring of acetylacetonate ligand according to the scheme presented in Figure 13.

Investigation of thermal decomposition of $Ir(CO)_2(acac)$ vapour in hydrogen showed that the mechanism of decomposition remains almost unchanged, only the concentration of acetylacetone in the gas phase slightly increases. In the presence of oxygen, H_2O is observed to appear in the gaseous reaction products.

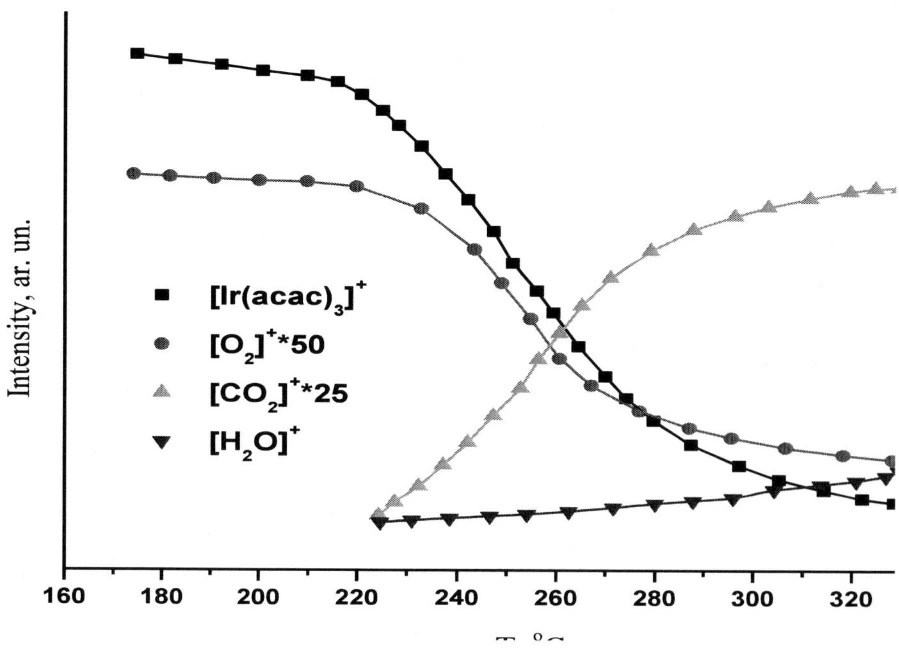

Figure 11. Temperature dependencies of the intensities of ion peaks in the mass spectrum of major gaseous products of thermolysis of $Ir(acac)_3$ vapour in the presence of oxygen.

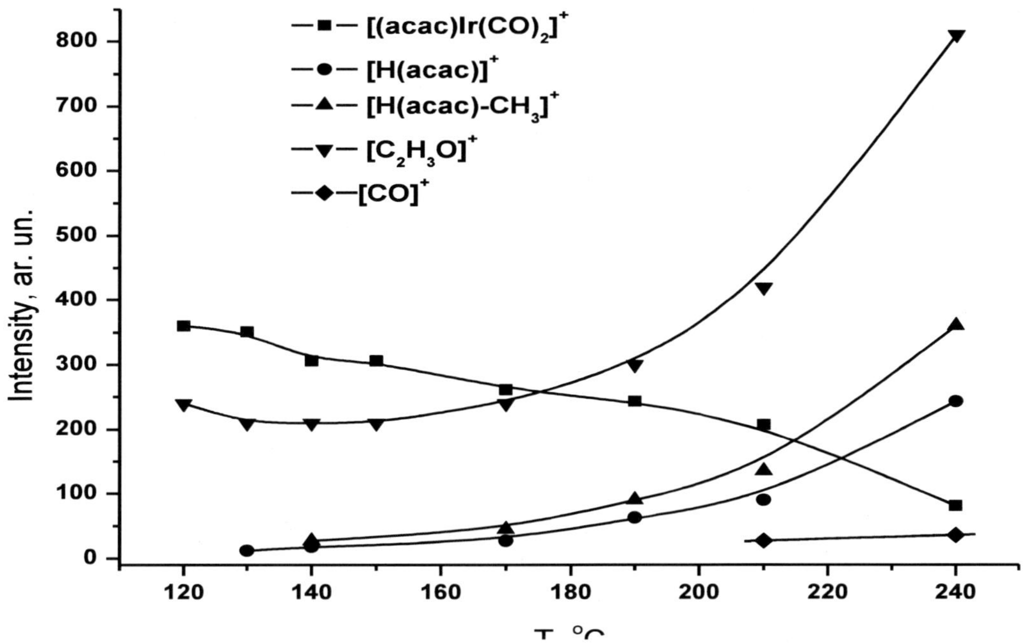

Figure 12. Temperature dependencies of the intensities of ion peaks in the mass spectrum of the major gaseous products during thermal decomposition of $(Ir(CO)_2(acac)$ vapour in vacuum.

The process of thermal decomposition of Ir(cod)(acac) vapour in vacuum (10^{-4} torr) starts at a temperature of 350°C and proceeds, similarly to the case of $Ir(CO)_2(acac)$, according to the bimolecular mechanism with the detachment of the neutral ligand – cyclooctadiene, and opening of the chelate ring of acetylacetonate ligand. Gaseous fragments of ligands can serve as additional sources of the carbon contamination of the films obtained using the heteroligand Ir(I) complexes.

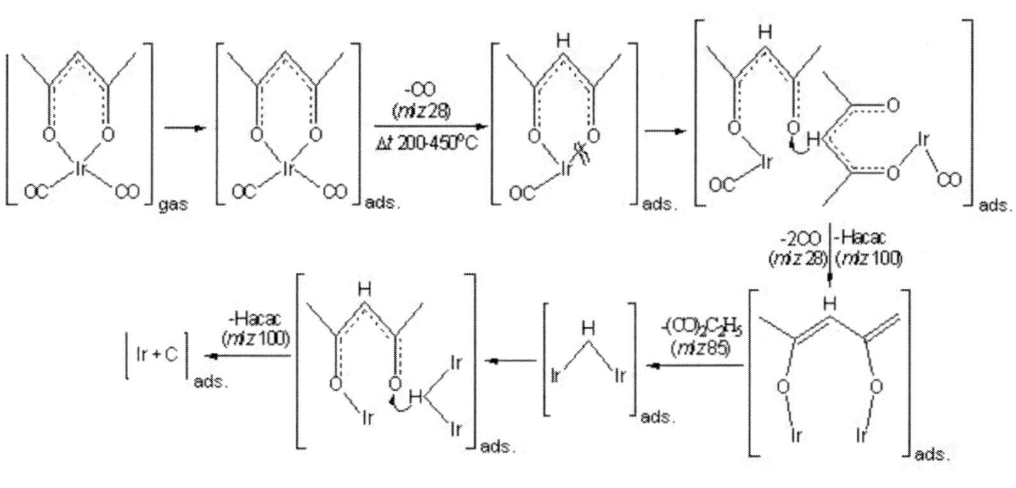

Figure 13. Scheme of possible routes of chemical transformations of $Ir(CO)_2(acac)$ vapour on heated surface in vacuum.

Deposition of Iridium Layers by Means of MOCVD

The thermal properties of the compounds in the condensed and gas phases, including thermal stability in the solid phase, vapour pressure and the routes of precursor vapour decomposition with (or without) reaction additives, determine the specific composition of the material of coating, its structure and morphology for the realization of chemical vapour deposition processes (MOCVD). The major requirements to the initial compounds can be formulated as follows:

- thermal stability of the precursor in the condensed state at vaporization temperature;
- high vapour pressure (volatility) of precursor;
- stability in the gas phase at vaporization temperature,
- decomposition of precursor vapour with the formation of the material of coating and gaseous products only;
- moderate temperature of film deposition, that is, not high threshold of precursor vapour stability;
- inertness of precursor vapour and gaseous decomposition products to the substrate material;
- non-toxic character, stability in the air, storage stability, high yield from synthesis, availability of starting reagents.

Many classes of the volatile compounds of iridium are used in MOCVD processes. The reviews [1, 7, 21] give a detailed presentation of halide, homo- and heteroligand, carbonyl, β-diketonate, cyclopenta- and cyclooctadienyl, alkyl, allyl, carboxylate derivatives of iridium, as well as the data on the available thermal characteristics of the compounds, deposition parameters, and a number of characteristics of the resulting layers. One of the problems of deposition remains the question how to obtain pure films without any admixtures of carbon or other elements. It may be stressed that any information about the application of the complexes with various sulphur-containing ligands as precursors in MOCVD processes is completely absent from literature, though the complexes of metals with sulphur-containing ligands are rather close to β-diketonate derivatives in thermal parameters, synthetic availability and storage stability.

The authors of given review carried out integrated investigations on obtaining iridium layers and studying their constitution [66-79].

By means of TEM, the structures of Ir films obtained from $Ir(acac)_3$ in the presence of H_2 at the initial stages of growth were investigated [71, 76, 77]. For thin discontinuous films (Figure 14), twinning and polytwinning along the (1 1 1) planes of the fcc lattice of Ir is characteristic, with the formation of secondary insular structures. Parallel twins shaped as flat blocks piled with (1 1 1) planes on each other are prevailing. For more disperse particles formed at low deposition temperatures, another kind of twinning occurs – stellate.

Films deposited at different temperatures differ from each other in the character if aggregation of primary structures. At a temperature of 370-420°C, Ir nanocrystallites 5-10 nm in size are aggregated into secondary isles 50-100 nm in size. Temperature rise leads to the crystallization of Ir into larger crystallites 20-50 nm in size, though the particles with smaller size are also present. At the deposition temperature of 500-550°C, ordered dendrite structures with the cross size of 50 nm are more characteristic. Such a character of the growth of Ir layers can be explained by the presence of an insignificant amount of co-deposited carbon-containing products (CP) formed during the thermal decomposition of $Ir(acac)_3$. Indeed, a thin amorphous layer 1-2 nm thick was detected on the surface of crystallites in the films obtained at a temperature above 470°C (Figure 14, *d*). No layers of this kind are formed at lower temperatures. As carbon is almost insoluble in Ir within a broad temperature range, it may be expected that in this case CP segregate at the interblock boundaries of crystallites and prevent Ir aggregation into coarse particles. With an increase in temperature, simultaneously with Ir crystallization, diffusion of CP to the surface of isles occurs, with the formation of a thin layer. Further growth of Ir films leads to the formation of continuous layers (Figure 15) formed as a result of accumulation and interaction of insular formations. For continuous films, the statistically established average size of Ir crystallites increases from 5 to 50 nm with an increase in deposition temperature from 370 to 550°C.

Thermal stability of Ir films was studied *in-situ* by annealing the films in a high-temperature micro-attachment of the electron microscope. Annealing of the film deposited at a temperature of 370°C caused sintering of microblocks 10 nm in size within 10 min at 700°C to form crystallites 50 nm in size without distortion of the film continuity. Similar annealing of the film deposited at 500°C did not cause any visible changes of its structure.

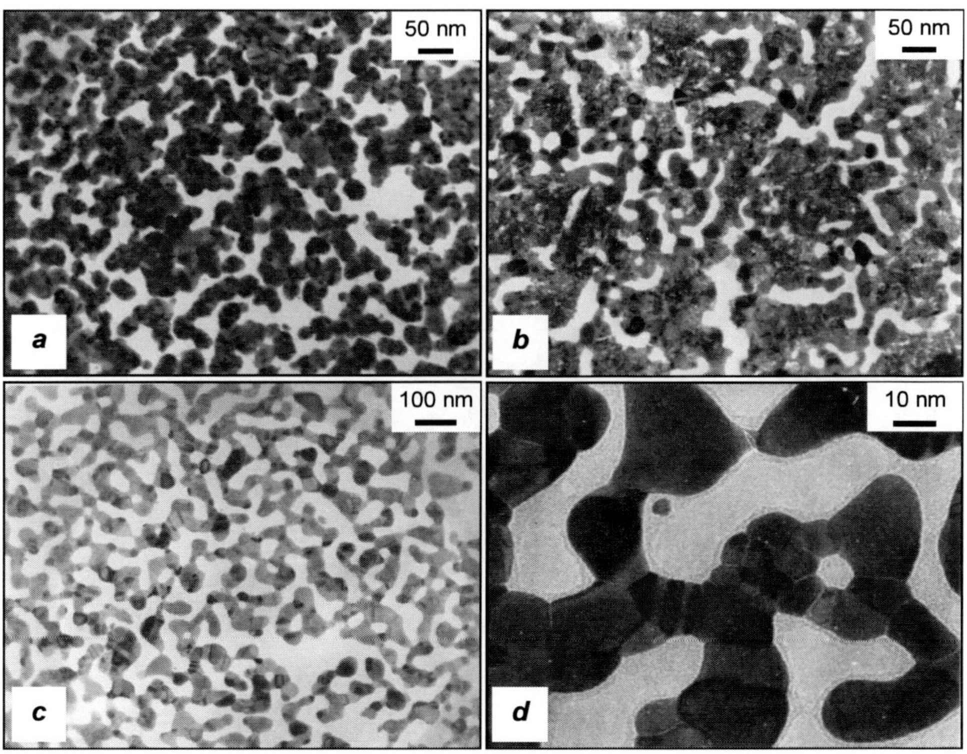

Figure 14. The structures of discontinuous Ir films obtained on quartz surfaces in the presence of H_2 at different deposition temperatures: (*a*) – 370°C, (*b*) – 420°C, (*c*) – 530°C; and amorphous layer (1-2 nm thick) of carbon-containing products of $Ir(acac)_3$ decomposition on the crystallite surface in Ir films deposited at a temperature of 470°C and higher (*d*).

To reveal the nature of particles forming Ir layers and to study their size distribution, non-destructive small-angle X-ray scattering technique (SAXS) was used for the first time [70, 76]. The Ir films were deposited onto the external surfaces of thin-walled quartz capillaries because the patterns are obtained in SAXS procedure when the X-ray beam passes directly through the sample. Unlike for wide-angle patterns, small-angle X-ray patterns contain the structural data on the disperse composition and particle size of not only crystallites but also other objects – non-uniformities of electron density (pores, defects etc.).

On the basis of the analysis of SAXS intensities on scattering angle, it may be supposed that the films are composed mainly of two classes of shapes: «coarsely homoaxial» or «compact» particles (with the axes ratio not less than 1:3) and «flattened» ones. On the basis of TEM data, we can suppose that the particles having flattened shapes in the structures of Ir films are flat crystallites, narrow gaps between the faces of crystallites, or thin films of carbon-containing fragments of the thermal decomposition of $Ir(acac)_3$ vapour. With an increase in deposition temperature, the size range of compact particles displays the trend to shift monotonously and broaden from $R \approx$ (4-11) nm at 370°C to $R \approx$ (10-28) nm at 550°C. Temperature change has a substantial effect also on other structural characteristics of the films (Table 7).

Coincidence of the results obtained in the investigation of disperse composition, shapes of particles and other structural parameters revealed by means of SAXS and by means of TEM provided a consistent notion of the internal microstructure of thin Ir films.

The phase composition of Ir films was studied by means of electron microdiffraction (Figure 15) [71, 76]. The major phase is metal Ir giving narrow intense diffraction lines (d_n: 2.22, 1.92, 1.35, 1.16 Å etc.). In addition, the traces of an additional phase are detected (d_n: 3.60, 2.55, 2.3, 1.62, 1.50, 1.25 Å). Examination of the films in the light-field and dark-field modes showed that this phase is represented in the structures of the films by discrete crystallites with the size comparable with the size of Ir crystallites. The content of this phase increases with an increase in deposition temperature.

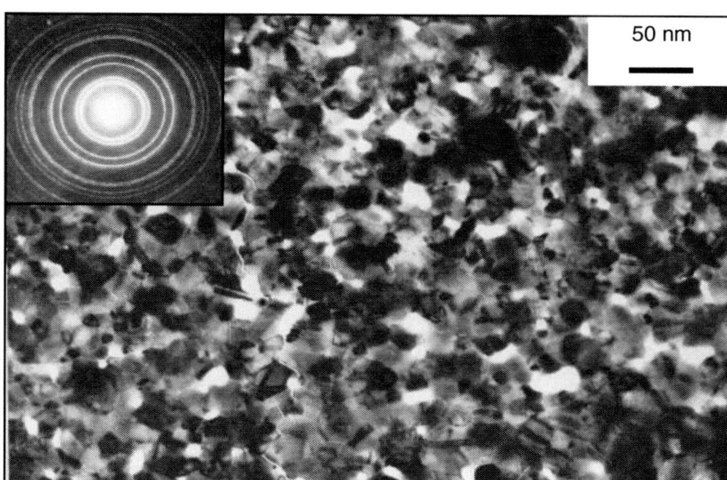

Figure 15. Typical polycrystalline structure of the continuous Ir film deposited on SiO_2 in the presence of H_2 at 500°C, and electron diffraction patterns from this film.

Table 7. Parameters of the structure of Ir films deposited on the surface of quartz capillaries at different substrate temperatures, determined on the basis of SAXS data

$T_{dep.}$, °C	R_g, Å	L_{max}, Å	S/V, m^2/mm^3	C_1	C_2	r, Å
370	74±3	220±10	0.34±0.02	1.0	-	-
420	104±5	320±15	0.42±0.09	0.51	0.49	30±2.5
470	135±7	410±15	0.29±0.03	0.35	0.65	48±2.0
500	161±10	455±25	0.21±0.03	0.68	0.32	10±1.0
530	155±9	450±20	0.16±0.01	0.97	0.03	13±1.5
550	168±11	460±20	0.16±0.01	0.94	0.06	12±1.5

T_{dep} – deposition temperature, R_g – average radius of gyration of the particles, L_{max} – maximal distance in the particles, S/V – specific surface of the film material, C_1 and C_2 – volume fractions of the particles having «compact» and «flattened» shapes, r – thickness of the flattened particles.

The chemical structure of Ir films was studied by means of X-ray photoelectron spectroscopy (XPS) with layer-by-layer analysis of the samples [69, 73, 76]. TEM examination of Ir layers obtained at the initial stages of growth on quartz substrates revealed the traces of an additional phase in the structure of the films. In this connection, special attention was paid to the analysis of the transient layer Ir–substrate.

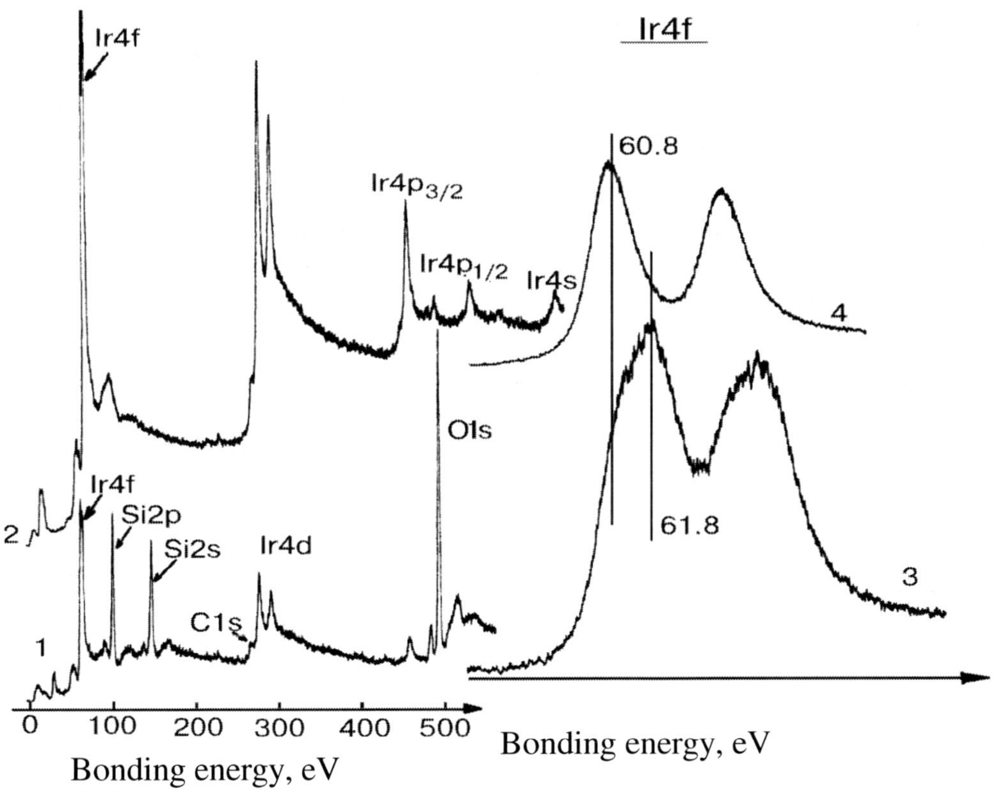

Figure 16. XP spectra of Ir film deposited at a temperature of 425°C on the quartz substrate in the presence of H_2. 1,2 – overview spectra, 3,4 – spectra of Ir4f; 2,4 – spectra of initial surface, 1,3 – spectra of the transient layer Ir–SiO_2.

Typical spectra characterizing Ir films deposited on SiO_2 in the presence of H_2 are shown in Figure 16. Curves 1, 2 correspond to the overview spectra, curves 3 and 4 – to the spectra of Ir4f. Spectra 2 and 4 were recorded after the purification of the surface by means of bombardment with argon ions. In addition to Ir lines, the overview spectrum 2 records a small amount of carbon and oxygen. A part of C-, O-containing products gets deposited together with Ir, while the major amount of impurities is due to the adsorption of CO on Ir surface during recording the spectra. According to the Ir4f spectrum (*curve* 4), Ir atoms are in the metal state. All the spectra recorded during ion etching of the film are identical to spectra 2 and 4. While approaching the Ir–SiO_2 interface, a decrease in the intensity of lines from Ir is observed, and the appearance of the lines related to the material of the substrate – Si2p, Si2s and O1s (*curve* 1). In this case, according to the spectrum of Ir4f (*curve* 3), Ir atoms are present in two non-equivalent states: metal, with binding energy 60,8 eV, and ionic, with binding energy 61,8 eV. Changes in binding energy were also observed in the spectra of Si2s, Si2p and O1s.

On the basis of the analysis of the entire set of XPS data (established correlation of chemical shifts for Ir4f, Si2s and O1s lines, as well as the synchronous character of a decrease in the intensities of Ir lines and the appearance of the lines related to the material of the substrate) it may be reliably accepted that the formation of compounds like $IrSi_xO_y$ occurs in the transient layer (Table 8). The formation of the «silicate» phase in the transition layer occurs for the deposition temperature 400°C and higher; at lower temperature only the metal state of Ir was observed. With an increase in deposition temperature, the relative thickness of the transient layer increases.

Investigation of Ir films deposited on SiO_2 in the presence of O_2 (Table 8) allowed revealing the effect of oxidative or reducing atmosphere on the sample composition. The formation of silicate phases in the region of the transient layer is characteristic also for these systems. Similarly, Ir is formed only in the metal state during further growth. Transient layers were detected also for the deposition of Ir on the substrates made of Al_2O_3, Si and Cu. Both for the deposition in the atmosphere of H_2 and in the presence of O_2, the thickness of the transient layer on oxide substrates and on silicon increases with an increase in temperature. Passing from the reductive to oxidative atmosphere, the temperature of the formation of transient layer decreases, while its thickness increases. For Cu as the substrate material in MOCVD of Ir, insertion of Cu atoms into the Ir film is observed; the plots of the dependence of the Ir/Cu ratio on the extent of ion etching of samples have a cupola-like character. Increased Cu content on the sample surface is explained by segregation of Cu into the oxide state during the interaction with the atmosphere.

The procedures for the deposition of Pd and Ir contacts on voluminous and film samples of high temperature superconducting (HTSC) ceramics having the composition $YBa_2Cu_3O_{7-x}$ were developed in [68]. The precursors were $Pd(hfac)_2$ and $Ir(acac)_3$. The temperature dependence of the contact electric resistance (**R**$_c$) for Pd and Ir layers was studied within the range 4-300 K. Near the temperature point of superconductor transition (**T**$_c$) dependence $R_c(T)$ has the semiconductor-type character for Pd contacts deposited in the presence of H_2. Such an appearance of $R_c(T)$ is the evidence of the presence of a boundary layer between the metal and the superconductor. An increase in the deposition temperature of Pd contacts causes an increase in the thickness of the transient layer.

Similar problems arose also for the deposition of Ir on the surface of HTSC ceramics in the atmosphere of H_2. In this connection, in order to obtain low-resistance Ir contacts, it was proposed to involve the deposition of Ir films in the presence of O_2. The deposition of Ir was performed within temperature range 300-400°C at atmospheric pressure. Within this temperature range, the surface of HTSC material gets free from water, and at the same time enrichment of the surface layer with oxygen occurs, which improves its structure.

For comparison, Table 9 shows the results of measurements of the resistance of contacts Pd – $YBa_2Cu_3O_{7-x}$, deposited in the presence of H_2, Ag (paste) – $YBa_2Cu_3O_{7-x}$ (after annealing) and Ir – $YBa_2Cu_3O_{7-x}$, deposited in the presence of O_2 (before and after annealing). One can see in the presented results that the resistance of Ir contacts is several orders of magnitude lower than that of Pd.

Additional thermal treatment causes a decrease in ρ_c by more than order of magnitude. It can be also stressed that for $T<T_c$ Ir samples have the metal-type conductance. The resistance of Ir contacts obtained in the presence of O_2 reaches the value acceptable for technical use. The contacts are deposited at a temperature of 300-400°C, that is, such a process can be used in combination with silicon-based technologies.

The pulse-MOCVD method was used to deposit ultra-thin Ir metal films from precursors with different types of ligands: $Ir(acac)_3$, $Ir(CO)_2(acac)$ on SiO_2/Si and Ta/Si substrates [66, 67]. Film deposition was performed in the set-up equipped with the systems of pulsed dosing of precursor vapour and co-reactant gas.

Table 8. Conditions of the formation and features of the chemical composition of Ir films

Substrate material	Reagent gas	Deposition temperature, °C	Temperature of the formation of transient layer, °C	Composition of transient layer
SiO_2	H_2	350-550	≥ 400	Ir – $IrSi_xO_y$
	O_2	280-420	≥ 360	
Al_2O_3	O_2	280-380	≥ 340	Ir – $IrAl_xO_y$
Si	H_2	300-550	≥ 350	Ir – Si (Ir_xSi_y)
	O_2	280-400	≥ 300	
Cu	H_2	250-550	≥ 250	Ir – Cu (Ir_xCu_y)

Table 9. Contact resistivity values $\rho_c = R_c \times S_c$ for systems [(Pd, Ir, Ag) – $YBa_2Cu_3O_{7-x}$]

Metal	Deposition temperature, °C	Reagent gas	Annealing temperature, °C	ρ_c(77 K), Ohm×cm^2
Pd	280	II_2	-	$7{,}5\times10^{-2}$
Pd	310	H_2	-	$4{,}2\times10^{-1}$
Ir	300	O_2	-	$2{,}2\times10^{-5}$
Ir	360	O_2	-	$1{,}6\times10^{-5}$
Ir	360	O_2	930	$1{,}4\times10^{-6}$
Ag	-	-	930	$2{,}5\times10^{-6}$

S_c – the area of contact surface.

Deposition of Ir films from Ir(acac)$_3$ was carried out by means of thermal- and pulse-MOCVD with presence H_2 or O_2 as gas-reactant [66].

The presence of light unsaturated hydrocarbons according to the data of mass spectrometric analysis points to the possibility of the formation of CP on the surface of the growing Ir film as a result of hydrogenation-dehydrogenation processes. Granular Ir films ~ 15 nm thick were deposited in vacuum (1×10^{-3} torr) in the mode of thermal MOCVD (substrate temperature: 470°C). The size of crystallites was 5-10 nm, the cavities between the Ir crystallites were filled with CP.

The addition of H_2 into the reaction zone does not lead to any changes of the mechanism of thermal decomposition of Ir(acac)$_3$. The yield of H(acac) increases, while the concentrations of other gaseous organic products decreases. In addition, in the presence of H_2 the thermal stability of H(acac) molecules increases up to 500°C. With further temperature rise, decomposition of H(acac) occurs, which causes contamination of the films; gas-phase decomposition of the precursor is also possible. By means of pulsed MOCVD with H_2 as the co-reactant gas, Ir films were obtained at the deposition temperature of 350-530°C (pressure in the reactor: ~ 10^{-2} torr, number of cycles: 1000-2500, thickness: 5-15 nm, growth rate: 0,04-0,15 Å/cycle). The typical structure of films obtained at reduced deposition temperature is shown in Figure 17 (*a*). The structure of the film is composed of Ir grains 3-5 nm in size, surrounded by CP.

Thus, investigation of the composition and structure of Ir films deposited by means of thermal and pulsed MOCVD in vacuum and in the atmosphere of H_2 confirms the assumptions made when studying thermolysis of Ir(acac)$_3$ vapour concerning possible co-deposition of CP.

Having analyzed the results of the investigation of Ir film deposition, we may conclude that the formation of granular structures during growth is determined by the formation of carbon-containing shells preventing coalescence of Ir particles.

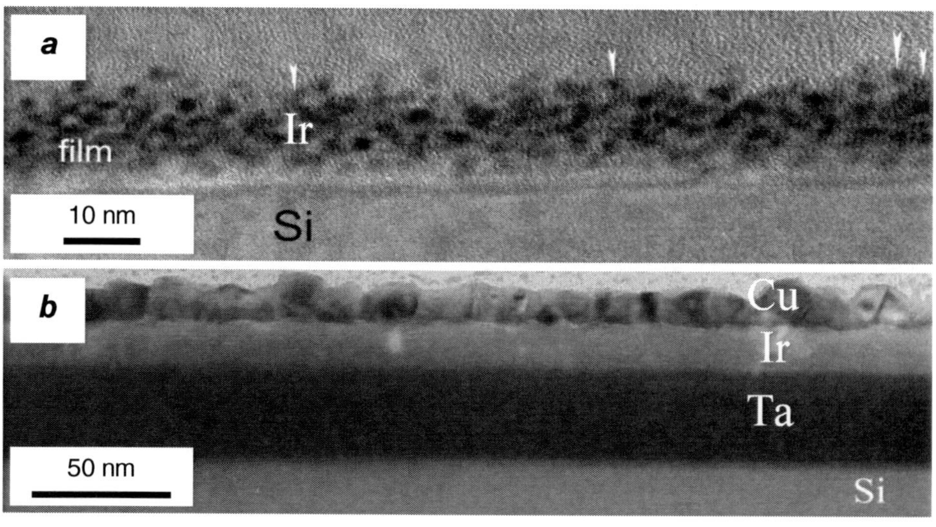

Figure 17. TEM images of the cross section of Ir films deposited from Ir(acac)$_3$ by means of pulsed MOCVD in the presence of H_2 at a temperature of 360°C (*a*) and in the presence of O_2 at a temperature of 310°C (Cu layer was deposited on Ir surface) (*b*).

According to the data obtained by means of mass spectrometry, the reactions involved in the interaction of $Ir(acac)_3$ with O_2 lead to the evolution of only CO, CO_2, H_2O and a very small amount of alcohols into the gas phase. With this character of thermal decomposition of $Ir(acac)_3$, one may expect that the amount of CP in the films will be minimal. With O_2 as the co-reactant gas, substrate temperature was varied within the range 300-370°C. At the deposition temperature of 300 and 310°C, Ir films ~ 20 nm thick were obtained (growth rate: ~ 0,2 Å/cycle). A typical structure of the films is shown in Figure 17 (*b*) (a thin germinal Cu layer was deposited on the surface of Ir film by means of pulse-MOCVD without decapsulation of the set-up, for subsequent galvanic additional growth). The Ir film has a continuous compact (non-granular) structure, it is practically uniform, and its minimal thickness is determined only by the roughness of the substrate (less than nanometer).

A set of studies on the deposition of Ir using $Ir(CO)_2(acac)$ was carried out [67]. The compound decomposes in vacuum and in H_2 with the formation of the following major products: H(acac), CO, CO_2 and C_2H_3O. In addition to the listed products, thermal decomposition of $Ir(CO)_2(acac)$ in the presence of O_2 is also accompanied by the formation of H_2O vapour. In this connection, it may be assumed that the composition and structure of Ir films would not be substantially dependent on the reactive additives used. Indeed, experiments in H_2 and O_2 at the deposition temperature of 240-300°C resulted in obtaining granular films with the average crystallite height of about 7 and 9 nm (according to the AFM data), respectively.

In [72], the samples of iridium coatings up to 50 μm thick on Mo substrates were obtained. The size of the characteristic elements of the surface of coatings decreases with a decrease in the rate of mass transfer of precursor vapor ($Ir(acac)_3$), the total content of gas-forming impurities in metal layers does not exceed $2{,}5\times10^{-2}$ mass %. For obtaining Ir coatings in the presence of oxygen, the amount of carbon contamination decreases in comparison with the process involving hydrogen, but the concentration of oxygen in the composition of the coating increases.

Investigation of the structure of sample cross section showed that under definite conditions the formation of columnar structures is typical for Ir coatings [77-79]. The appearance of columnar structures was observed on the substrates of different nature (metals, oxides). At the initial stages of growth, the coating with compact structure is deposited, then the growth mechanism changes and the layer with columnar structure is formed (Figure 18).

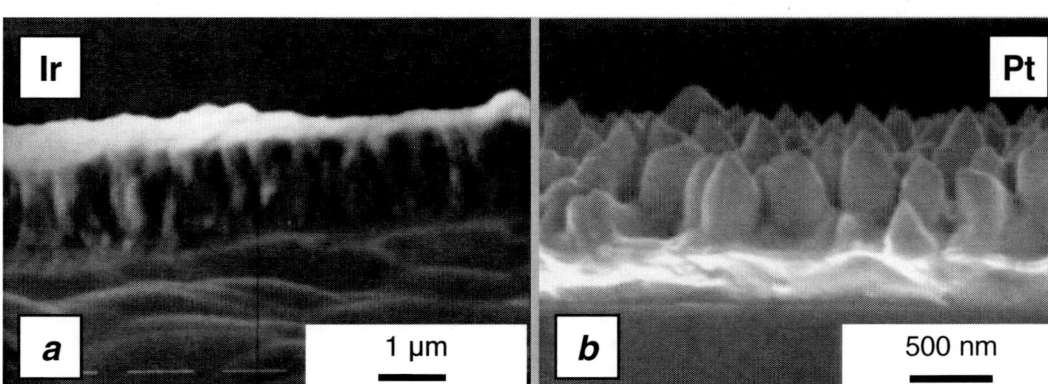

Figure 18. Columnar structures: Ir coating (a), Pd coating (b).

For Pt deposition as example (Figure 18, *b*), it was demonstrated that a thin layer with high carbon concentration is formed on the surface of the growing metal, which leads to structural changes as the process goes on, and results in the formation of coatings with non-uniform layered structure [77].

In [78, 79], the methods for the deposition of Ir and composite Ir–Al_2O_3 coatings from metal acetylacetonates on steel substrates coated preliminarily with Al_2O_3 layer were developed. Deposition temperature, evaporator temperature and the composition of the gas phase were varied in the experiments.

Independently of deposition parameters, Ir coatings deposited in H_2 cracked or partly separated from the substrate when rapidly cooled, which is likely due to the absence of the transient layer in Ir–Al_2O_3 at the deposition temperature of 450°C. Doping additive Al_2O_3 was introduced to change the mechanical properties of the coating. Two major approaches were implemented: preparation of the coatings with gradient composition, with the maximal Al_2O_3 content at the boundary with the substrate, and preparation of mixed Ir–Al_2O_3 coatings with approximately uniform distribution of the doping additive over the thickness. In both cases the coatings did not get separated from the substrate. In the first case the lower layer has a compact structure, and the upper Ir layer is columnar. In the second case the structure of the coating changed substantially: the coating was composed of several layers of compact continuous structure. A detailed analysis of the composition of the upper layers demonstrated that the mixed coating is composed of the layers of metal Ir separated by thinner layers composed of Al_2O_3 and metal Ir.

In the presence of O_2, Ir coatings were deposited at the deposition temperature of 320 and 340°C. Temperature rise by 20°C leads to the coatings with loose discontinuous structure, to an increase in the size of Ir crystallites from 6 to 25 nm and to an increase in the concentration of dissolved oxygen in the composition. At the deposition temperature of 320°C, the rate of coating growth was varied by changing the concentration of precursor vapour in the deposition zone. Ir layer with the most dense structure was obtained with the growth rate of ~ 3-4 μm/h, an increase in the growth rate to ~ 8 μm/h resulted in the formation of the coating composed of the layers with different structures (compact, columnar and granular) (Figure 19, *a*).

The most uniform continuous structure is observed in the mixed Ir–Al_2O_3 coatings with the ratio Ir/Al ~ 2,2 (Figure 19, *b*). The orientation mainly manifested in the sample is [1 1 1]. The coating is composed of Ir nanocrystallites and X-ray amorphous Al_2O_3. With the introduction of the oxide phase into the coating, the size of Ir crystallites decreased by a factor of 2 (from 6 to 3 nm). It may be assumed that Al_2O_3 affects the process of Ir crystallization either by the formation of the nanoparticles of the oxide phase or through the formation of amorphous layers on the surface of Ir crystallites thus preventing an increase in their size.

In [74, 75], investigation of the corrosion stability of coatings based on platinum group metals (Ir, Pt, Ir-Pt, Ru) on Ti electrodes was carried out. The coatings were obtained from metal acetylacetonates at atmospheric pressure in the presence of H_2 (Ir) or O_2 (Ir, Pt, Ir-Pt, Ru), deposition temperature and the method of electrode preparation were also varied (substrates were annealed before the deposition of coatings and electrodes with Ru coatings).

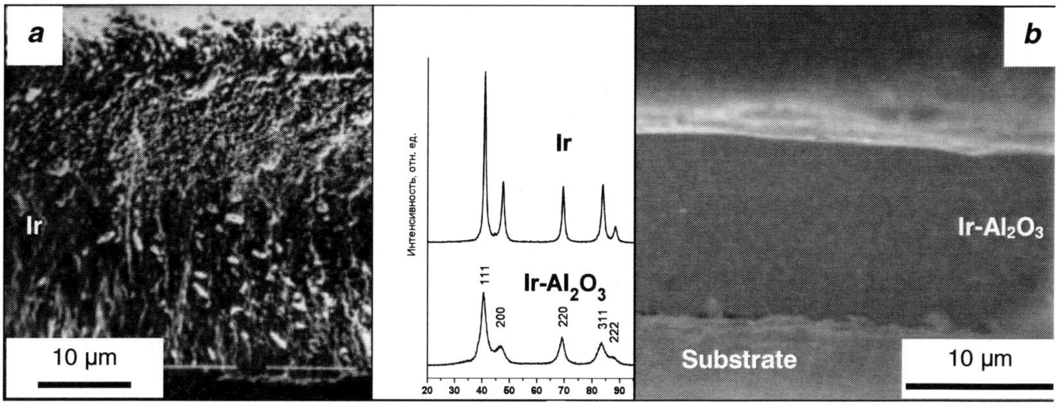

Figure 19. SEM images of the chip of Ir (*a*) and Ir-Al_2O_3 (*b*) coatings deposited in the presence of O_2, and diffraction patterns from these samples.

Investigation of the corrosion stability of the samples was performed with the electrolysis of seawater imitate (NaCl 31 g/l, Na_2SO_4 3 g/l). Anodes were the samples under investigation. Counter-electrodes were Ti plates. The anode current density was 960 A/m^2. The composition of the solution and electrolysis conditions (static electrolyte, pH=9.2±0.5, active evolution of O_2) were chosen according to the principle of the severe action on electrodes to decrease the duration of experiments. For the same purpose, relatively thin coatings were deposited (0,2-1,7 μm).

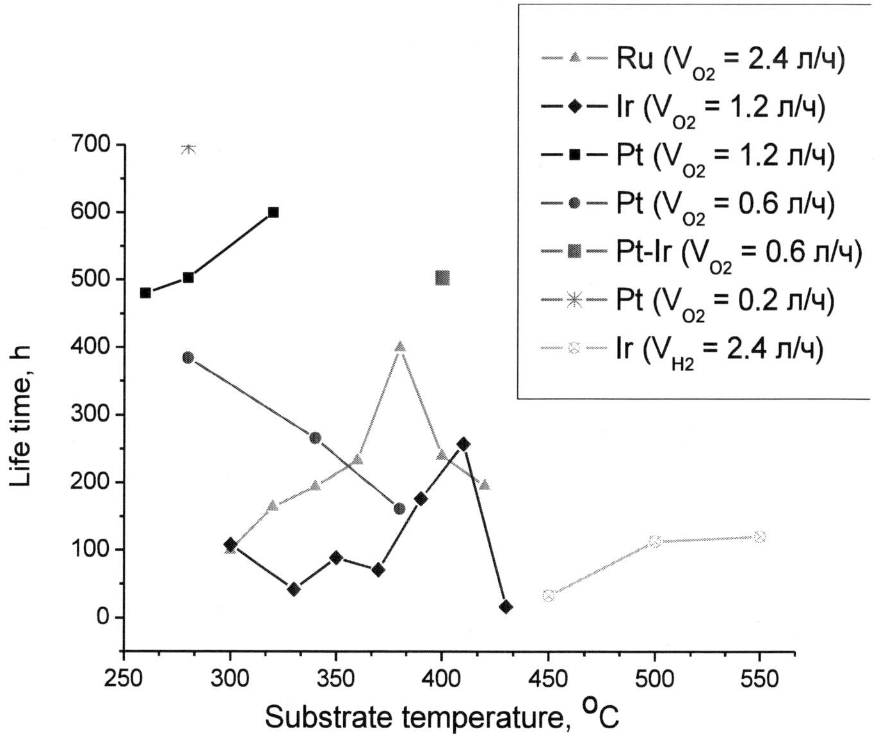

Figure 20. Dependence of the corrosion stability of samples on deposition temperature.

The time of sample performance is essentially dependent on the composition of coatings and the conditions of their preparation (Figure 20). For instance, the stability of Ir coatings obtained in the presence of O_2 increases substantially with an increase in the substrate temperature from 300 to 410°C (from 42 to 257 hours). At the same time, an increase in deposition temperature by 20°C more (to 430°C) causes a sharp worsening of the situation. In this case, thermal decomposition of $Ir(acac)_3$ occurs in the gas phase but not on the surface, so the coating is fine Ir powder. The stability of Ir layers obtained in the atmosphere of H_2 also increases with an increase in substrate temperature but remains smaller than that of high-temperature coatings deposited in the presence of O_2.

An interesting result was obtained for the mixed Pt-Ir coating. This sample is a composite based on oxide phases of Pt and Ir, and the solid solution of these metals. The corrosion stability of the Pt-Ir layer turned out to be much higher than that of Ir and Pt coatings obtained in the similar deposition mode.

The external manifestation of the failure of a sample was either detachment of a substantial part of the coating or passivation of the electrode (that is, a strong increase in potential, above 3 V), even if the external continuity appears to be conserved. The first type of behavior is characteristic of the samples with Ir and Ru coatings, and it is also observed with the electrode coated with Pt-Ir layer. In the case of Ir coatings the observed phenomenon is connected with a sharp difference in the coefficients of thermal expansion and elastic modulus of Ti and Ir, which leads to the formation of cracks and to weak adhesion of the coating to the substrate. It is also connected with the wedging action of water molecules during the hydration of the interface between the substrate and the boundary (the so-called Rebinder effect). The second type of behavior is exhibited by the anodes with Pt layers. An electrode with the coating is a complicated system: metal (Ti)–semiconductor (solid solution of titanium oxide in Ti)–platinum group metal (or conducting oxide). Diffusion doping of Ti oxide by the material of the coating is possible during the deposition of the coating, which improves the conductivity of the contact between the electrode and the coating. In addition, the conductivity of such a contact can be also due the injection of electrons from the coating into the oxide layer on Ti by means of tunneling transition. A probable mechanism of conductivity worsening at the contact between the electrode and the coating may be hydration of titanium oxide due to porosity and permeability of the coating, which results in an increase in the thickness of the interface and a decrease of the probability of tunneling injection of electrons through it.

To test the working capacity of Ti electrodes with MOCVD coatings, an electrodialysis set-up intended for additional purification of water entering through the pipelines of Gorvodokanal (Novosibirsk, Russia) was built. The anode current density was 100-200 A/m^2. The electrodes were made of Ti foil (thickness: 400 μm, $S = 8{,}75\times10^{-2}\ m^2$) with Ir coating. In order to decrease the testing time, the conditions for the deposition of the least corrosion-proof sample were chosen for manufacturing the electrodes with Ir coating. The set-up successfully operated for more than 4000 hours; no visible deterioration of the coatings was detected. Thus, the quantitative data on the corrosion stability of coatings during electrolysis were obtained under the conditions of severe action on the samples, which allows one to evaluate the working capacity of protective layers based on platinum group metals for various electrochemical applications.

CONCLUSION

So, in the present chapter we analyzed the available data on the synthesis and investigation of the thermal properties of volatile iridium complexes with organic ligands, the principles for the selection of initial compounds for the implementation of MOCVD processes, and a number of results obtained by the authors on the deposition and examination of iridium films and coatings.

Chemistry of volatile metal-organic compounds of platinum group metals, in particular iridium, is now at the new stage of its development due to the appearance of new possible applications of such complexes in different areas, involving formation of a nanostructured thin solid film materials by means of Metal-Organic Chemical Vapor Deposition (MOCVD). So, this chapter was focused on:

- systematization of the known experimental data on the synthesis of iridium volatile compounds with the following ligands: beta-diketonates, carbonyls, dienes, cyclopentadienyls, allyls, as well as heteroligand complexes; and also the description of both conventional and original procedures developed by the authors;
- the data on the thermal properties of these complexes in the condensed phase, obtained by means of complex thermal analysis;
- - the data on the measurements of temperature dependencies of saturated vapor pressure performed using tensimetry (flow method, Knudsen method, static method with membrane null manometer), as well as calculations of thermodynamic parameters of vaporization processes;
- results of the investigation of thermal transformations of the compound vapour on the heated surface, gaseous by-products and mechanisms of thermal decomposition processes, with the description of unique procedures developed by the authors with colleagues using *in situ* high-temperature mass spectrometry;
- data on deposition and characterization of Ir-containing films and coatings. Determination of correlations between deposition conditions, composition and structure of films.

The synthesis of Ir precursors with required characteristics, study of precursor thermal properties and development of Ir MOCVD processes remain one of the actual tasks at solving the problems of obtaining films and coatings for different application.

REFERENCES

[1] *Application of metal-organic compounds for preparation of inorganic coatings and materials* / Ed. by. G.A. Razuvaev. - Moscow: Science, 1986. - 256 p.

[2] Hoke J.B., Stern E.W., Murray H.H. Low-temperature vapour deposition of high-purity coatings from cyclooctadiene complexes of iridium // *J. Mater. Chem.* - 1991. - V. 1, N 4. - P. 551-554.

[3] Rubezhov A.Z. Platinum group organometallics // *Platinum Metals Rev.* - 1992. - V. 36, N 1. - P. 26-33.

[4] Cohan J.S., Yuan H., Williams R.S., Zink J.I. Laser-assisted organometallic chemical vapor deposition of films of rhodium and iridium // *Appl. Phis. Lett.* - 1992. - V. 60, N 11. - P. 1402-1403.

[5] Hamilton J.C., Yang N.Y.C., Clift W.M., Boehme D.R., McCarty K.F., Franklin J.E. Diffusion mechanism in chemical vapor-deposited iridium coated on chemical vapor-deposited rhenium // *Met. Trans.* - 1992. - V. 23A, N 3. - P. 851-855.

[6] Goto T., Vargas R., Hirai T. Preparation of iridium and platinum films by MOCVD and their properties // *Journal de Physique IV.* - 1993. - V. 3. - P. 297-304.

[7] *The chemistry of metal CVD* / Ed. by T. Kodas, M. Hampden-Smith. - Weinhem; New York; Basel; Cambridge; Tokyo: VCH, 1994. - 534 p.

[8] Vargas R., Goto T., Zhang W., Hirai T. Epitaxial growth of iridium and platinum films on sapphire by metalorganic chemical vapor deposition // *Appl. Phys. Lett.* - 1994. - V. 65, N 9. - P. 1094-1096.

[9] Goto T., Vargas R., Hirai T. Preparation of iridium clusters by MOCVD and their electrochemical properties. // *Mater. Sci. Eng.* - 1996. - V. A217/218. - P. 223-226.

[10] Goto T., Ono T., Hirai T. Microstructure and electrical properties of Ir-C composite films prepared by MOCVD. // *Funtai oyobi Funmatsu* Yakin. - 1998. - V. 45, N 1.- P. 20-24.

[11] Sweeney C.E., Sun Ya.-M., Ekerdt J., White J. M. CVD Ir film growth and vapor pressure analysis with iridium tris-acetylacetonate. *National Organization for Black Chemists and Chemical Engineers. Proc.* – NOBCChE. - 1998. - N 25. - P. 131-136.

[12] Tuffias R. H. Fabrication processes for iridium/rhenium combustion chambers // *Mater. Manuf. Processes.* - 1998. - V. 13, N 5. - P. 773-782.

[13] Sun Y. M., Endle J. P., Smith K., Whaly S., Mahaffy R., Ekerdt J. G., White J. M. and Hance R. L. Iridium film growth with iridium tris-acetylacetonate: oxygen and substrate effects. // *Thin Solid Films.* - 1999. - V. 346, N 1-2. - P. 100-107.

[14] Goto T., Vargas R. and Hirai T. Effect of oxygen gas addition on preparation of iridium and platinum films by metal-organic chemical vapor deposition. Mater. // *Trans., JIM.* - 1999. - V. 40, N 3. - P. 209-213.

[15] Goto T., Hirai T., Ono T. Ir cluster films prepared by MOCVD as an electrode for a gas sensor. // *Trans. Mater. Res. Soc.Jpn.* - 2000. - V. 25, N 1. - P. 225-228.

[16] Goto T., Ono T., Hirai T. Electrochemical properties of amorphous carbon/nano-granular iridium films prepared by metalorganic chemical vapor deposition (MOCVD). // *Funtai oyobi Funmatsu Yakin.* - 2000. - V. 47, N 4. - P. 386-390.

[17] Sun Y.-M., Yan X.-M., Mettlach N., Endle J. P., Kirsch P. D., Ekerdt J. G., Madhukar S., Hance R. L., White J. M. Precursor chemistry and film growth with (methylcyclopentadienyl)(1,5-cyclooctadiene)iridium. // *J. Vac. Sci. Technol.* - 2000. - A. - V. 18, N 1. - P. 10-16.

[18] Endle J.P., Sun Y.-M., Nguyen N., Madhukar S., Hance R.L., White J.M., Ekerdt J.G. Iridium precursor pyrolysis and oxidation reactions and direct liquid injection chemical vapor deposition of iridium films. // *Thin Solid Films.* - 2001. - V. 388, N 1-2. - P. 126-133.

[19] Goto T.; Ono T.; Hirai T. Electrochemical properties of iridium-carbon nano composite films prepared by MOCVD. // *Scr. Mater.* - 2001. - V. 44, N 8-9. - P. 1187-1190.

[20] Leskela M., Ritala M. Atomic layer deposition (ALD): from precursors to thin film structures // *Thin Solid Films.* - 2002. - V. 409. - P. 138-146.

[21] Garcia J.R.V., Goto T. Chemical vapor deposition of iridium, platinum, rhodium and palladium. // *Materials Transactions.* - 2003, - V. 44, N 9. - P. 1717-1728.

[22] Aaltonen T, Ritala M, Tung YL, Chi Y, Arstila K, Meinander K, Leskela M. Atomic layer deposition of noble metals: Exploration of the low limit of the deposition temperature // *J. Materials Research.* - 2004. - V. 19, N 11. - P. 3353-3358.

[23] Aaltonen T, Ritala M, Sammelselg V, Leskela M. Atomic layer deposition of iridium thin films // *J. Electrochem. Soc.* - 2004. - V. 151, N 8. - P. G489-G492.

[24] Hua Y.F., Zhang L.T., Cheng L.F., Yang W.B. Structural and morphological characterization of iridium coatings grown by MOCVD // *Materials Science and Engineering B.* - 2005. - V. 121, Is. 1-2. - P. 156-159.

[25] Kimura T., Suzuki G., Goto T. Preparation of noble metal (Ru, Ir)-carbon nano-composites by MOCVD as catalytic electrode for oxygen sensors. // *Journal of Metastable and Nanocrystalline Materials.* - 2005. - V. 24, N. 25. - P. 589-592.

[26] Hu C., Wan J., Dai J. Iridium/Carbon films prepared by MOCVD. *Platinum Metals Rev.* - 2005. - V. 49, N 2. - P. 70-76.

[27] Leskela M., Aaltonen T., Hamalainen J., Niskanen A., Ritala M. Atomic Layer Deposition of Metal Thin Films // *Proceedings of the Fifteenth International European Conference on Chemical Vapor Deposition* (EUROCVD-15). Bochum, Germany, 2005. The Electrochemical Society, Inc., Pennington, NJ, USA, Proceedings Volume 2005-09 / Editors: A. Devi, R. Fischer, H. Parala, M.D. Allendorf, M.L. Hitchman. - P. 545-554.

[28] Yang W.B., Zhang L.T., Cheng L.F., Hua Y.F., Zhang J. Structural and morphological characterization of iridium coatings grown by MOCVD // *Rare metal materials and engineering.* - 2006. - V. 35, Is. 3. - P. 488-491.

[29] Chen Y.-L., Hsu C.-C., Song Yi-H., Chi Y., Carty A.J., Peng S.-M., Lee G.-H. Iridium metal thin films and patterned IrO2 nanowires deposited using iridium(I) carbonyl precursors // *Chem. Vap. Deposition.* - 2006. - V. 12, N 7. - P. 442-447.

[30] Ritterhaus Y., Hur'yeva T., Lisker M., Burte E.P. Iridium thin films deposited by liquid delivery MOCVD using Ir(EtCp)(1,5-COD) with toluene solvent. *Chem. Vap. Deposition.* - 2007. - V.13, N 12. - P. 698-704.

[31] Ritterhaus Y., Hur'yeva T., Lisker M., Burte E.P. Iridium electrodes for ferroelectric capacitors deposited by liquid-delivery MOCVD and PVD // *ECS Transactions.* - 2007. - V. 2, N 7. - P. 67-78.

[32] Ohriner E. K. Processing of iridium and iridium alloys, Methods from purification to fabrication // *Platinum Metals Review.* - 2008. - V. 52, N 3. - P. 186-197.

[33] Kim S.-W., Kwon S.-H., Kang S.-W. Two step annealing of iridium thin films prepared by plasma-enhanced Atomic Layer Deposition // *ECS Transactions.* - 2008. - V. 16, N 4. - P. 309-314.

[34] Yang W., Zhang L., Hua Yu., Cheng L. Thermal stability of iridium coating prepared by MOCVD // *Journal of Refractory Metals & Hard Materials.* - 2009. - V. 27. - P. 33-36.

[35] Bonati F., Ugo R. Iridium Dicarbonyl β-Diketonates // *J. Organometal. Chem.* - 1968. - V. 11. - P. 341-352.

[36] Jesse A.C., Meester A.M., Stufkens D.J., Vrieze K. Vibrational and NMR Spectroscopic Studies on (acac)M(substituted olefin) // *Inorgan. Chimica Acta.* - 1978. - V. 26. - P. 129-136.

[37] Hay-Motherwell R.S., Wilkinson G. Synthesis and Reactions of Tris (tetramethylethylenediaminelithium) Hexamethylmetallates of Rhodium, Iridium and Ruthenium(III) and of Hexakis(neopentyl)(μ-oxo)Dirhodium(IV) // *Polyhedron.* - 1990. - V.9, N 17. - P. 2071-2080.

[38] Dwjer F.P., Sargeson A.M. The preparation of tris-acetylacetone-rhodium(III) and iridium(III) // JACS.- 1953. - V. 75, N 4. - P.984-985.

[39] Bennett M.A., Mitchel T.R.B. γ-C-bonded 2,4-pentadionato complexes of trivalent iridium // *Inorg. Chem.* - 1976. - V. 15, N 11. - P. 2936-2938.

[40] Stossel P., Bash I. et al. *Method for Producing Iridium(III) ketonates.* Merck Patent GMBH. WO 2006/018202 A1. Feb. 23, 2006.

[41] Masayuki S., Takemoto S. Raw Material Compound for CVD, and Method of Vacuum Chemical Vapor Deposition for Iridium or Iridium Compound Thin Film. *Patent Abstracts of Japan.* 2003-321416. Nov. 11, 2003.

[42] Bentler P., Gamsjager H. Preparation and ultraviolet visiblespectrum of hexa-aquairidium(III) // *J. Chem. Soc. Chem. Com.* - 1976. - N. 14. - P. 554-555.

[43] Isakova V.G., Igumenov I.K., Zemskov S.V. *The method of preparation of Ir(III) β-diketonates.* Inventors Certificate N797226. Nov. 11, 1980 (in Russia).

[44] Igumenov I.K., Shipachev V.A., Gorneva G.A., Morozova N.B. *The method of preparation of Rh(III) and Ir(III) β-diketonates.* Patent N2105008. Feb. 20, 1998 (in Russia).

[45] Igumenov I.K., Isakova V.G., Morozova N.B., Shipachev V.A. *The method of preparation of rare metal tris-β-diketonates.* Patent EA N000402. April 09, 1999 (in Russia).

[46] Zherikova K.V., Kuratieva N.V., Baidina I.A., Morozova N.B. Crystal structure of *trans*-Ir(tfac)$_3$ // *Zhurn. of Structur. Khimii.* - 2010. - V. 51, N 4. - P. 796-799 (in Russia).

[47] Igumenov I.K., Zharkova G.I., Isakova V.G., Zemskov S.V. β-Diketonates of precious metals / in book "Theoretical and applied chemistry of metal β-diketonates" Moscow: Science, 1985. - P. 36-51.

[48] Igumenov I.K., Zharkova G.I., Isakova V.G., Semyannikov P.P., Grankin V.M., Zemskov S.V. Volatile β-diketonates of precious metals / *Proceedings "Precious metals. Chemistry and Thechnology".* Novosibirsk. 1989, P. 186-205.

[49] Isakova V.G., Baidina I.A., Morozova N.B., Igumenov I.K. γ-Halogenated iridium (III) acetylacetonates // *Polyhedron.*-2000.-N19.-P.1097-1103.

[50] Davignon L., Dereigne A., Bonnaire R., Manoli J.M. Complexes β-Dicetonato-(π-Cyclooctadiene-1,5)Iridium. Syntheses et Etude Cristallographique // *J. Less-Common Metals.*-1971.-V.25.-P.75-81.

[51] *The methods of elemento-organic chemistry. Types of transition metal organic compounds* / by ed Nesmeyanov A.N. and Koshechkov K.A. Moscow: Science. Book 2. 1975, 947 p.

[52] Ball R.G., Graham A.G., Heinekey D.M. et al. Synthesis and Structure of [(ή-C_5Me_5)Ir(CO)]$_2$ // *Inorg. Chem.* - 1990. - V. 29. - P. 2023-2025.

[53] Seiji O., Shunichi F, Kiyoshi I. *Method for Producting Dihalogenocyclopentadienyliridium dimmer.* Patent Abstracts of Japan. 2005-281169. Oct. 13, 2005.

[54] Hirotosai U., Katsumi A.M. *Organic Iridium Compound.* Patent Abstracts of Japan. 08-306627. Nov. 11, 1996.

[55] Morozova N.B., Semyannikov P.P., Trubin S.V., Stabnikov P.P., Bessonov A.A., Zherikova K.V., Igumenov I.K. Vapor pressure of some volatile iridium(I) compounds with carbonyl, acetylacetonate and cyclopentadienyl ligands // *J. of Therm. Anal. and Cal.* - 2009. - V. 96(1). - P. 261–266.

[56] John K.D., Salazar K.V., Scott B.I., Baker R.T. et al. Chemistry of M(allyl)$_3$ (M=Rh, Ir) compounds: structural characterization of tris(allyl)iridium complexes with phosphorus ligands // *Chem. Commun.* - 2000. - P. 581-582.

[57] Clark T., Robinson S.D. Complexes of the Platinum Metals. Part 43. N,N'-Diphenylamidinato Derivatives of Ruthenium, Osmium and Iridium // *J. Chem. Soc. Dalton Trans.* - 1993. - V. 1. - P. 2827-3502.

[58] Sysoev S.V., Morozova N.B., Zharkova G.I., Igumenov I.K., Semyannikov P.P., Grankin V.M. Vapour pressure and thermoanalytical study of diethyldithiocarbamates of platinum metals // *J. of Therm. Anal.* - 1998. - V. 53. - P. 87-96.

[59] Raston C.L., White A.H. Crystal structure of tris(diethyldithiobycarbamato)-iridium(III) // *J. Chem. Soc. Dalton Trans.* - 1976. - P. 32-35.

[60] Logvinenko V.A. Thermal analysis of coordinated compounds and clathrates / Novosibirsk: Science. *Siberian Branch.* 1982, 128 p.

[61] Kumar S., Kaushik N.K. Synthesis, Characterization & Thermal Studies on Ru(III), Rh(III), Ir(III), Pd(II), Pt(II) & Os(IV) Dithiocarbamates // *Indian J. Chem.* - 1981. - V. 20A. - P. 512-515.

[62] Suvorov A.V. *Thermodynamic Chemistry of vaporous state* / Leningrad: Chemistry. 1970, 206 p.

[63] Jesse A.C., Baks A., Stufkens J., Vrieze K. Thermochemical Studies on (Acac)M(Olefin)$_2$ (M = Rh(I), Ir(I)) // *Inorg. Chim. Acta.* - 1978. - V. 29. - P. 177-182.

[64] Semyannikov P.P., Igumenov I.K., Trubin S.V., Asanov I.P. *In situ* mass spectrometry during thermal CVD of the *tris*-acetylacetinates of 3-d transition metal // *Journal de Physique IV.*-2001.-V.11.-P.Pr3-995–Pr3-1003.

[65] Isakova V.G., Semyannikov P.P., Grankin V.M., Igumenov I.K. Thermal stability of Ir(III) and Rh(III) β-diketonates // *Koord. Khim.*- 1988. - V. 14, N 1. - P. 57-62.

[66] Gelfond N.V., Semyannikov P.P., Trubin S.V., Morozova N.B., Igumenov I.K. Deposition of Ir Nanostructured Thin Films by Pulse CVD // *ECS Transactions.* - 2009. - V. 25, N 8. - P. 871-874.

[67] Semyannikov P.P., Morozova N.B., Zherikova K.V., Trubin S.V., Igumenov I.K., Gelfond N.V. Preparation of Iridium Thin Film by Pulse CVD from Ir(acac(CO)$_2$) // *ECS Transactions.* - 2009. - V. 25, N 8. - P. 887-892.

[68] Bessergenev V.G., Gelfond N.V., Igumenov I.K., Ilyasov S.Sh., Kangiev R.D., Kovalevskaya Yu.A., Kravchenko V.S., Slobodyan S.A., Motorin V.I., Shestak A.F. Technology of fabrication and electrical characteristics of W, Mo, Ni and Ir, Pd contacts with $YBa_2Cu_3O_{7-x}$ // *Supercond. Sci. Technol.* - 1991. - V. 4, N 7. - P. 273-278.

[69] Gelfond N.V., Igumenov I.K., Boronin A.I., Bukhtiyarov V.I., Smirnov M.Yu., Prosvirin I.P., Kwon R.I. An XPS study of the composition of iridium films obtained by MO CVD // *Surface Science.* - 1992. - V. 275. - P. 323-331.

[70] Gelfond N.V., Tuzikov F.V., Igumenov I.K. Effect of the deposition temperature on the iridium film microstructure produced by metal-organic chemical vapour deposition: sample characterization using X-ray techniques // *Thin Solid Films*. - 1993. - V. 227, N 2. - P. 144-152.

[71] Gelfond N.V., Zaikovskii V.I., Igumenov I.K. Electro-microscopy study of Ir thin film structure deposited by MOCVD // *Surface. Physic, chemistry, mechanic*. - 1994. - N 5. - P. 50-57.

[72] Igumenov I.K., Gelfond N.V., Tuffias R.H. Thick platinum metal coatings obtained by MO CVD // *Chemical Vapor Deposition. Proceedings of the Fourteenth International Conference and EUROCVD-11*. Paris. - 1997. Electrochemical Society Proceedings Volume 97-25 / Editors: M.D. Allendorf, C. Bernard. - P. 1618-1625.

[73] Gelfond N.V., Boronin A.I., Smirnov M.Yu., Kvon R.I., Igumenov I.K. An XPS study of the influence of iridium CVD conditions on the film composition // *Chemical Vapor Deposition. Proceedings of the Fourteenth International Conference and EUROCVD-11. Paris*. - 1997. Electrochemical Society Proceedings Volume 97-25 / Editors: M.D. Allendorf, C. Bernard. - P. 1588-1595.

[74] Gelfond N.V., Galkin P.S., Igumenov I.K., Morozova N.B., Fedotova N.E., Zharkova G.I., Shubin Yu.V. *MO CVD obtaining composite coating from metal of platinum group on titanium electrodes* // J. de Physique IV. - 2001. - V. 11. - P. 593-599.

[75] Igumenov I.K., Gelfond N.V., Galkin P.S., Morozova N.B., Fedotova N.E., Zharkova G.I., Shipachev V.A., Reznikova E.F., Ryabtsev A.D., Kotsupalo N.P., Titarenko V.I., Dikov Yu.P., Distler V.V., Buleev M.I. Corrosion testing of platinum metals CVD coated titanium anodes in seawater simulated solutions // *Desalination*. - 2001. - V. 136. - P. 273-280.

[76] Badalyan A.M., Belyui V.I., Gelfond N.V., Igumenov I.K., Kosinova M.L., Morozova N.B., Rastorguev A.A., Rymyantsev Yu.M., Smirnova T.P., Fainer N.I., Yakovkina L.V. Chemical texture and structure of thin films deposited by MOCVD // *Zhurn. Structur. Khimii*. - 2002. - V. 43, N 4. - P. 605-628.

[77] Igumenov I.K., Gelfond N.V., Morozova N.B., Nizard H. Overview of coating growth mechanisms in MOCVD processes as observed in Pt group metals // *Chemical Vapor Deposition*. - 2007. -V. 13, N 11. - P. 633-637.

[78] Gelfond N.V., Morozova N.B., Igumenov I.K., Filatov E.S., Gromilov S.A., Kvon R.I. Structure of Ir-Al_2O_3 coatings, deposited by MOCVD in presence of hydrogen // *Zhurn. Structur. Khimii*. - 2009. - V. 50, N 5. – P. 955-959.

[79] Gelfond N.V., Morozova N.B., Igumenov I.K., Filatov E.S., Shubin Yu. V., Gromilov S.A., Kvon R.I., Danilovich V.S. Structure of Ir and Ir-Al_2O_3 coatings, deposited by MOCVD in presence of oxygen // *Zhurn. Structur. Khimii*. - 2010. - V. 51, N 1. - P. 88-97.

In: Iridium: Occurrence, Characteristics and Applications ISBN: 978-1- 62081-681-3
Editors: Chieko Fukui and Maemi Ono

Chapter 2

RECENT ADVANCES OF HOMOGENEOUS IRIDIUM CATALYSTS IN AQUEOUS SOLUTION

Wan-Hui Wang and Yuichiro Himeda
Energy Technology Research Institute,
National Institute of Advanced Industrial Science and Technology,
Higashi, Tsukuba, Ibaraki, Japan

ABSTRACT

Compared to the widely studied rhodium and ruthenium complexes, the high potency of iridium complexesin organometallic catalysis has only recently been recognized. Within the last decade, in particular, the high activity of homogenous iridium catalysts has been demonstrated; thus, they have become increasingly important in both academic and industrial research. From anenergy and environmental perspective, the investigation of bio-inspired catalytic reactions in the ultimate green medium — water — is very intriguing for chemists. This chapter presents recent advances in iridium-catalyzed aqueous reactions including the interconversion of carbon dioxide/hydrogen and formic acid, water oxidation, and transfer hydrogenation. Iridium complexes have shown promising catalytic activity for interconversion between CO_2/H_2 and HCOOH, which has recently been determined to be an environmentally friendly method for hydrogen storage. Water oxidation is fundamentally important for alternative fuel production, while safe and non-toxic transfer hydrogenation using a green hydrogen donor is an excellent alternative to hydrogenation. The design, synthesis, and application of representative iridium complexes for these bio-inspired reactions, as well as their catalytic mechanisms, are covered.

1. INTRODUCTION

Remarkable advances have been made in homogeneous catalysis over the past 50 years. Among the many metals, rhodium, ruthenium, and palladium have played pivotal roles in both academic and industrial research [1,2]. There is no doubt that the majority of homogenous catalysts stem from the pioneering studies of homogenous hydrogenation. The discovery of Wilkinson's hydrogenation catalyst, $RhCl(PPh_3)_3$, wasan important milestone in organometallic chemistry. Interestingly, its iridium analogue, $IrCl(PPh_3)_3$, is entirely in active towards hydrogenation due to strong binding between iridium andthe phosphine ligand. In the 1960s, Vaska's complex,$IrCl(CO)(PPh_3)_3$, contributed significantly to the understanding of oxidative addition [3]. However, Vaska's complex showed poor catalyticactivity towards hydrogenation because the H_2 adduct, $[IrH_2Cl(CO)(PPh_3)_2]$, is stable; therefore, the required phosphine dissociation to providea site for the binding of the alkene substrate does not occur. Nevertheless, the iridium catalyst was stilluseful forthestudy of the catalytic mechanism since iridium usually forms strong metal-ligand bonds, which allows for the isolation of relevant intermediates. In1977, Crabtree reported a mixed-ligand complex, $[(cod)Ir(Py)(PCy_3)]^+$ (cod: cyclooctadiene; Py: pyridine; Cy:cyclohexyl), that allowed for the homogenous hydrogenation of hindered tri- and tetra-substituted alkenes [4]. The iridium complex surprisingly out performed its ruthenium and rhodium analogues in terms of catalytic activity and substrate scope. Since then, the development of homogenous iridium catalystshas dramatically increased. Various carefully designed iridium complexes with improved activity have been utilized in a variety of transformations besides hydrogenation.

Environmental and energy issues have recently become a major global problem. Many chemists are devoted to finding solutions to these issues. The activation of small molecules (e.g., H_2, CO_2, and H_2O, etc.) is crucial to the fields of energy and the environment. Effective interconversion between H_2/CO_2 and HCOOH will contribute to utilization of CO_2 and/or the storage of hydrogen. Catalysis of water-splitting to form H_2 and O_2 is a central element in artificial photosynthesis.

In addition, the development of aqueous catalysts is of great interest for green chemistry. Water is the preferred solvent because it is abundant, safe, clean, and inexpensive. Moreover, using water as a solvent has recently been found to improve the reaction rate and selectivity in some cases [5]. Remarkable advances in homogeneous iridium catalysts in these fields have been made in the last decade.

This chapter aims to draw attention to the activation of small molecules by homogenous iridium catalyst in the ultimate green solvent —water.

2. Hydrogenation of Carbon Dioxide

The enormous emissions of carbon dioxide from the burning of fossil fuels for energy significantly contribute togreenhouse gases and global warming. Since the industrial revolution, the concentration of CO_2 in the atmosphere has risen remarkably. A further increase in energy demand will result in increased emissions of CO_2. In nature, plants transform CO_2 into carbohydrates in an aqueous environment at ambient conditions to store energy. Inspired by this natural process, the reduction of inert CO_2 for energy storage has attracted considerable interest. Unfortunately, a suitable method for storing energy (e.g., in form of hydrogen) based on the reduction of CO_2 is currently unavailable. The development of technologies to efficiently store hydrogen in a safe and reversible manner is an important challenge in order to realize the hydrogen economy. Presently, intensive efforts are being devoted to the development of low-cost, large-scale storage of hydrogen for fuel cell applications. In the last four decades, the homogeneous catalytic hydrogenation of CO_2 to

HCOOH and its derivatives has been widely studied. There are some reviews related to this subject [5–11].

Recently, formic acid attracted attention as one of the most promising hydrogen storage materials (Figure 1) because of its low toxicity, easy storability, and transportability [12-14]. In this section, the homogeneous iridium-catalyzed hydrogenation of CO_2 into formic acid or formate for hydrogen storage will be presented. Among the homogeneous catalysts for the hydrogenation of CO_2, ruthenium- and rhodium-phosphine complexes have shown the highest catalytic activity. Although research into iridium catalystsdates back to the 1970s, its promising catalytic activity has only recently been discovered. In 1976, Inoue and co-workers first reported the hydrogenation of carbon dioxide to yield formic acid using a homogeneous iridium catalyst [15].

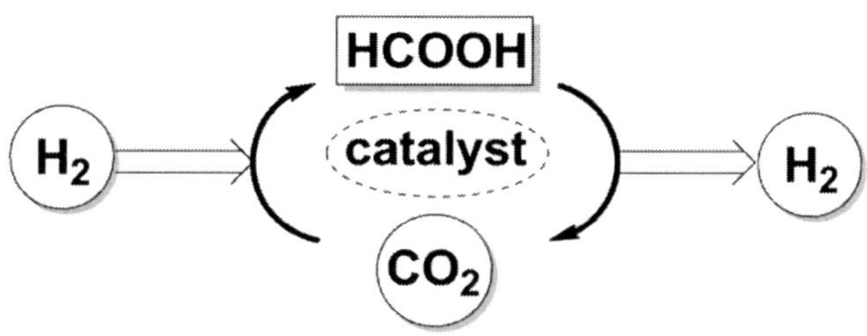

Figure 1. Interconversionbetween H_2/CO_2 and formic acid.

The reaction proceeded with CO_2 and H_2 (CO_2/H_2 = 1/1) under 5 MPa in benzene/H_2O using triethylamine as an additive. The yields of formic acid varied depending on the metal catalyst and phosphine ligands. $H_2Ru(PPh_3)_4$ exhibited a higher activity than $RhCl(PPh_3)_3$ and $H_3Ir(PPh_3)_3$, and resulted in the highest turnover number (TON) of 87.

About 20 years later,an iridium catalyst was again applied to CO_2 hydrogenation.[16] Rhodium(I) and ruthenium(II) complexes with a water-soluble phosphine ligand, 3-sulfonatophenyldiphenylphosphine (TPPMS), showed a turnover frequency (TOF) up to 262 h^{-1} in aqueous solution under mild conditions without the need of an organic co-solvent or amine additive. However, the iridium complex, $[IrCl(CO)(TPPMS)_2]$, showed no formate production under the same conditions.

In 2008, Gonsalvi and Laurenczyet al. reported a half-sandwich iridium complex, 1, bearing a water-soluble phosphine ligand, 1,3,5-triaza-7-phosphaadamantane (PTA). [17] Compared to the ruthenium and rhodium analogues [18], it gave a much lower TOF of 22.6 h^{-1} at 100°C. These preliminary studies implied that iridium complexes only provide low catalytic activity for the hydrogenation of CO_2. However, researchers recently made a break through and demonstrated the high catalytic ability of iridium complexes resulting in renewed attention on iridium complexes. The representative iridium catalysts, 1–9, are presented in Figure 2 and the catalytic results are listed in Table 1.

Despite numerous previous studies, several critical problems for practical applications remain to be solved. The first is the high activity of the catalyst, which is generaly required in order to lower the overall energy barrier for the conversion of thermodynamically stable CO_2. The second is the reusability of the catalyst, which is important to increase the cost efficiency. The third is tuning thecatalytic activity in both the reaction and separation steps to avoid potential reverse reactions. The final is the prevention of waste generation (e.g., organic solvents and additives) during the reaction.

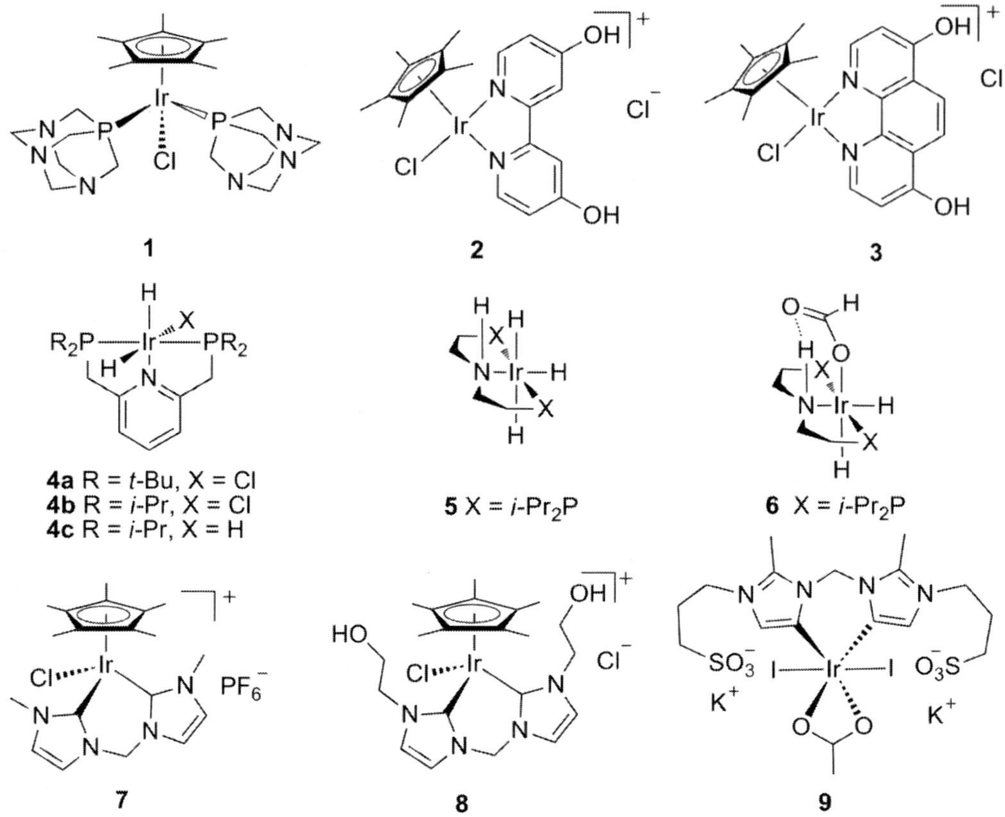

Figure 2. Representative iridium catalysts for the hydrogenation of CO_2.

Table 1. Hydrogenation of CO_2 to formic acid/formate using iridium catalysts

Catalyst	Solvent	Additive	T (°C)	P (MPa)	*t* (h)	Initial TOF (h^{-1})	TON	ref
$H_3Ir(PPh_3)_3$	C_6H_6/ H_2O	NEt_3	rt	5	20	0.8	15	[15]
1	H_2O	-	100	10	-	23	-	[17]
2	H_2O	KOH	120	6	57	42,000	190,000	[19]
2	H_2O	KOH	60	0.1	50	33	376	[19]
3	H_2O	KOH	60	0.1	50	32	444	[19]
3	H_2O	KOH	30	0.1	30	3.5	81	[19]
3	H_2O	KOH	120	6	48	33,000	222,000	[19]
4c	H_2O/ THF	KOH	200	5[a]	2	150,000	300,000	[20]
4c	H_2O/ THF	KOH	120	6[a]	48	73,000	3,500,000	[20]
6	H_2O	KOH	185	5.5	24	14,500	348,000	[21]
6	H_2O	KOH	185	5.5	1	18,780	18,780	[21]
7	H_2O	KOH	80	6	18	-	1600	[22]
8	H_2O	KOH	200	4	20	-	9500	[23]
9	H_2O	KOH	200	6	75	2,500	190,000	[24]

[a] Total pressure at room temperature.

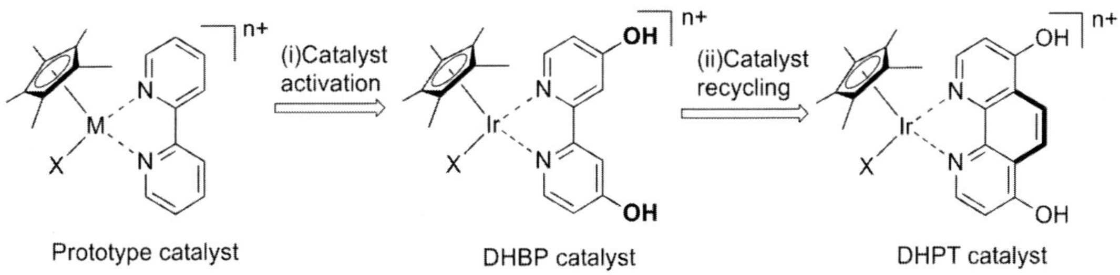

Figure 3. Design of the catalysts.

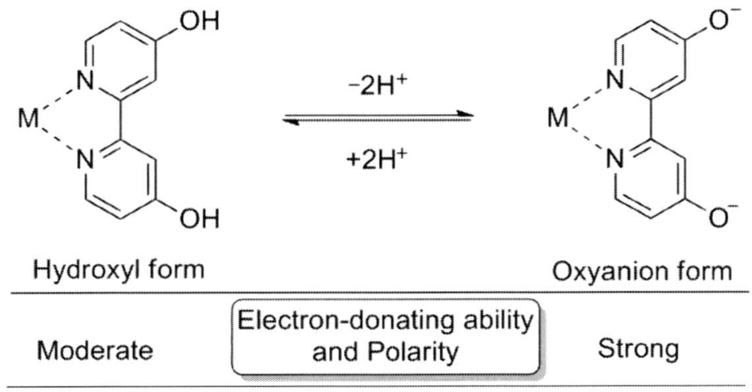

Figure 4. Acid-base equilibrium between hydroxyl and oxyanion forms.

(i) High catalytic activity was achieved by introducing two electron-donating hydroxyl groups onto the bipyridine ligand. More importantly, under basic conditions, the hydroxyl groups can be deprotonated to generate an oxyanion, which is a much stronger electron donor. Table 2 shows the effect of the addition of thehydroxyl group in the half-sandwich bipyridine catalyst, $[(C_nMe_n)M(L)Cl]^+$ (M = Rh, Ir, n = 5; M = Ru, n = 6), on the hydrogenation of CO_2. Significant activation of the catalysts was observed. The turnover numbers of the iridium catalysts with hydroxyl groups were 52–103 times greater than those of the unsubstituted catalysts. The electronic substituent effect was investigated using $[(C_nMe_n)M(4,4'\text{-}R_2\text{-}2,2'\text{-bpy})Cl]^+$ (M = Ir, Rh, Ru; R = OH, OMe, Me, H). Note that the hydroxyl group ($\sigma_p^+ = -0.92$) was deprotonated to generate an oxyanion ($\sigma_p^+ = -2.30$) under basic conditions. The Hammett plots show a good correlation between the initial TOFs and the σ_p^+ values (Figure 5). On the other hand, the substituent effects on the rhodium and ruthenium complexes were moderate compared to the effect on iridium complex 2 (Figure 5, a vs. b and c) [29]. It is apparent that the remarkable activation of the iridium DHBP catalyst can be attributed to the strong electron-donating ability of the oxyanion. The catalytic activity of the iridium DHBP catalyst was maximized (TOF = 42,000 h^{-1}, TON = 190,000) at 6 MPaand 120 °C. Furthermore, the reaction proceeded at atmospheric pressure. These results indicate that the corresponding hydride complex can easily be generated as an active species at atmospheric pressure.

(ii) Catalyst recyclingwas realized by replacing the DHBP ligand with 4,7-dihydroxyl-1,10-phenanthroline (DHPT) [25]. The DHPT catalyst, 3, showed a similar TOF and a slightly improved TON than DHBP catalyst 2 (Table 1). The abovementioned acid-base equilibrium not only changes the electronic properties of the complex but also affects its polarity and thus its water solubility. As shown in Figure 6, DHPT catalyst 3 exhibited negligible solubility (ca. 100 ppb) in a weakly acidic formate solution. Therefore, recycling of 3was investigated in batch-wise cycles based on the concept shown in Figure 7. When the added KOH was completely consumed, the catalyst precursor spontaneously precipitated due to its decreased watersolubility as a result of the lower pH value. Thus, the reaction was terminated and formed a heterogeneous system that could be filtered to recover the precipitated catalyst. The iridium complex remaining in the filtrate was found to be less than 2% of the catalyst loading (0.11 ppm). Since the catalytic action was "turned off", the reverse reaction (i.e., the decomposition of formic acid) was prevented in the separation step. Additionally, the pure product (i.e., the formate salt) could be isolated simply by evaporating the filtrate.

Table 2. Substituent effect of the ligand on the TON for hydrogenation of CO_2[a]

Cat.[b]		TON			
	L:	bpy	DHBP	phen	DHPT
Rh-L		216[c]	1,800	220	2,300
Ru-L		68	4,400	78[c]	5,100
Ir-L		105[c]	5,500	59	6,100

[a] The reaction was carried out with thecat. (0.1 mM) in a 1 M KOH solution under 4 MPa (CO_2/H_2 = 1:1) at 80 °C for 20 h. [b] Rh-L = [Cp*Rh(L)Cl]Cl, Ru-L = [(C6Me6)Ru(L)Cl]Cl, Ir-L = [Cp*Ir(L)Cl]Cl. c.[Cat.] = 0.2 mM.

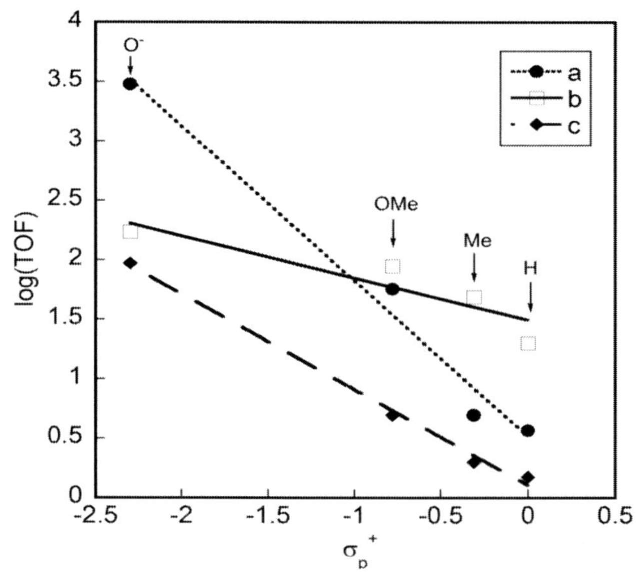

Figure 5. Correlation between initial TOFs and σ_p^+ values of substituents (R) for the hydrogenation of CO_2 catalyzed by $[(C_nMe_n)M(4,4'\text{-}R_2\text{-}2,2'\text{-bpy})Cl]Cl$. a) M = Ir, n = 5; b) M = Rh,n = 5; c) M = Ru,n = 6; R = OH, OMe, Me, H. The reactions were carried out in an aqueous1 M KOH solution under 4 MPa (CO_2:H_2 = 1:1) at 80 °C for 20 h.

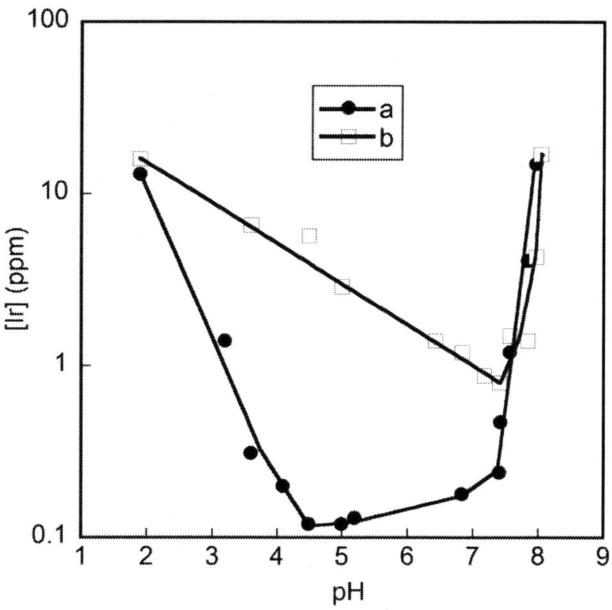

Figure 6. Solubility of a) DHPT catalyst 3and b) DHBP catalyst 2 in a 1 M aqueous formate solution.

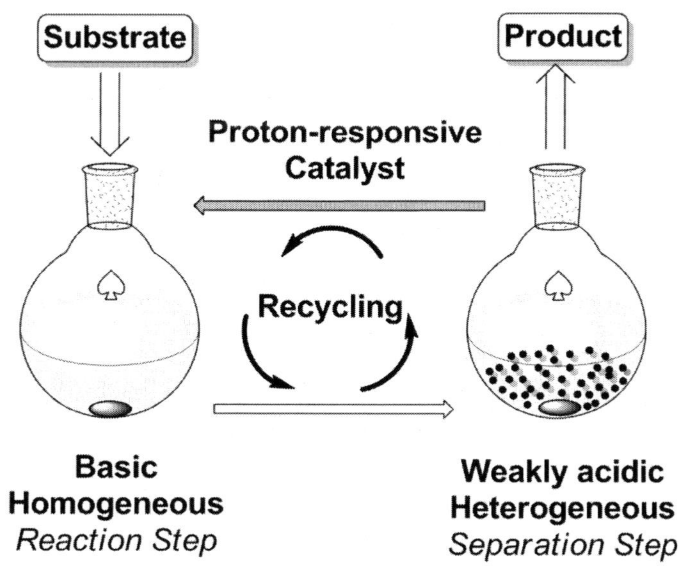

Figure 7. Recycling of proton-responsive catalyst with tunable solubility.

The recovered catalyst retained a high catalytic activity across four cycles, as shown in Table 3. It is clear that the three components (i.e., catalyst, product, and solvent) can be easily separated without significant waste. The sophisticated design of the catalyst provided aproton-responsive catalyst with pH-tunable catalytic activity (through electronic effects) and water solubility (through polarity). These results suggest that by carefully considering reaction profiles, the designand use of innovative homogeneous catalytic systemssuch as tailor-made catalysts can overcome the disadvantages of homogeneous catalysis.

Table 3. Catalyst recycling studies for the conversion of CO_2 into formate using iridium-DHPT catalyst 3[a]

Cycle	Loaded/ recovered cat.(ppm)	Recovery efficiency (%)	Leaching iridium (ppm) [b]	Final conc. of formate (M)
1	9.0	-	0.11	0.105
2	8.4	93	0.22	0.104
3	7.7	92	0.42	0.103
4	7.0	91	0.61	0.103

[a] Optimized conditions: DHPT catalyst (2.5 μmol), 6 MPa of H_2/CO_2 (1:1), 0.1 M KOH solution (50 mL), 60 °C for 2 h.

[b] Determined by ICP-MS analysis.

In 2009, Nozaki and co-workers designed Ir(III) complexes 4a–c in which alkylphosphine-based pincer ligands were employed as efficient electron donors. These complexes were used for the hydrogenation of CO_2 in H_2O/THF. The PNP-Ir trihydride complex, 4c, showed the highest TON (3,500,000) and TOF (150,000 h^{-1}) to date [20]. They also proposed a mechanism for the catalytic reaction (Figure 8): the insertion of CO_2 into 10 gives formato complex 11, which undergoes dissociation of the formato ligand under basic conditions. Simultaneously, deprotonative dearomatization of the PNP ligand by OH^- leads to intermediate 12, which is hydrogenated to regenerate the trihydride complex, 10.

Soon after, the mechanism was studied by Ahlquist et al. through a simple (PNP)IrH_3 structure using density functional theory (DFT) [30]. Their research suggested that the deprotonation by the hydroxide is the rate-limiting step (Figure 8).

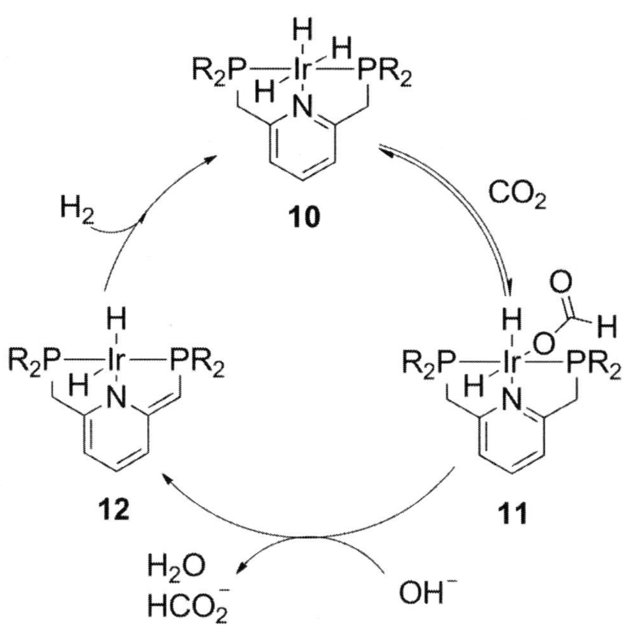

Figure 8. Catalytic mechanism for the hydrogenation of CO_2 proposed by Nozaki.

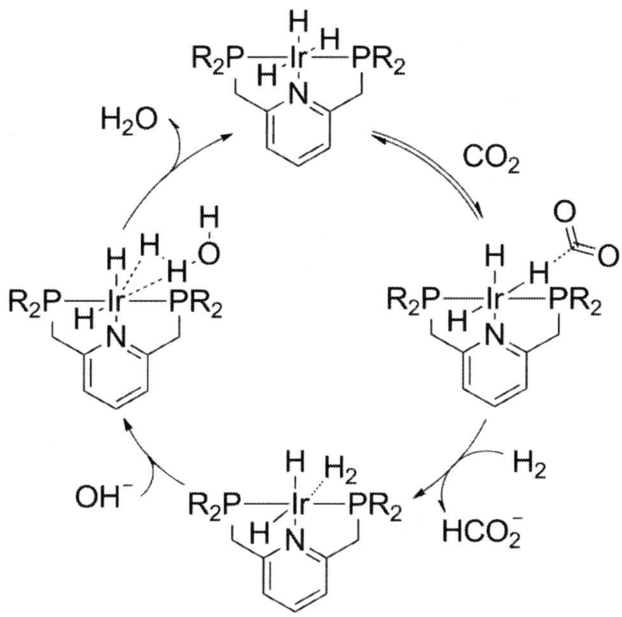

Figure 9. Catalytic mechanism proposed by Yang.

This calculation agreed with the experimental observation that higher basicity leads to a higher rate. However, this calculation could not identify intermediate 12 and gave a relatively high over all energy barrier of 109.3 kJ mol^{-1}. Most recently, Yang reinvestigated this mechanism using the DFT method and proposed a different reaction pathway (Figure 9) [31]. He suggested that direct H_2 cleavage by OH^- is more favorable than the Nozaki-postulated H_2 cleavage model. Using this new reaction pathway, the calculation gave a low overall enthalpy barrier of 77.8 kJ mol^{-1} for the formation of HCOOH from H_2 and CO_2.

In 2011, Hazari and co-workers investigated CO_2 insertion into PNP-Ir hydrides using a computational method [21]. They evaluated the nucleophilicity of the hydride through its calculated NBO chargeand found a strong correlation between the NBO charge of the hydride and the thermodynamics of CO_2 insertion. Using this simple model, they predicted that complex 5 is favorable for CO_2 insertion (Figure 10). Furthermore, they experimentally isolated air- and moisture-stable complex 6 and fully characterized its structure, which was confirmed by X-ray diffraction analysis. When 6 was used for the hydrogenation of CO_2, maximum TON and TOF of 348,000 and 18,780 h^{-1}, respectively, were obtained, which is comparable tothe best system reported by Nozaki. Via DFT calculations, they determined that the hydrogen bond in complex 6 facilitates CO_2 insertion. In the proposed mechanism, the rate-limiting step was suggested to be the displacement of formate by H_2 (Figure 10).

N-Heterocyclic carbenes (NHCs), which have a high electron-donating ability, have also been introduced as ligands iniridium complexes for the hydrogenation of CO_2. Most recently, Peris and co-workers reported a series of IrCp*(NHC) complexes. A bis-NHC Ir complex, 7, showed modest activity (TON of 1,600) for the hydrogenation of CO_2 to HCOOK [22]. To improve the water solubility of the complex, hydroxyl groups were introduced to the side carbon chains. Consequently, complex 8 gave a higher TON of 9,500 under optimized conditions [23].

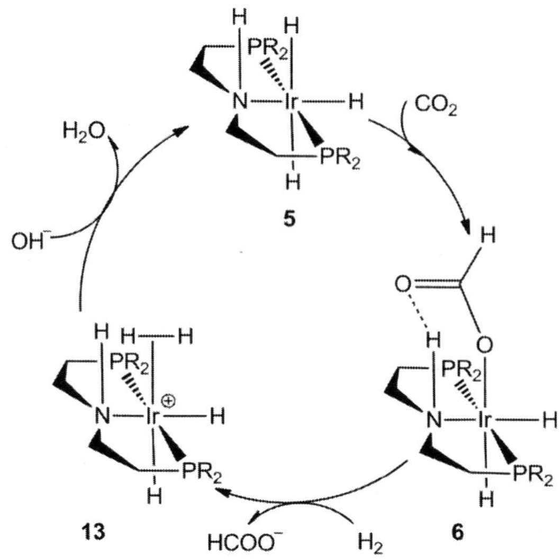

Figure 10. Reaction mechanism proposed by Hazari based on experimental and computational studies.

Furthermore, blocking the C2-position of imidazole with a methyl group and coordinating to the C5 position led to a higherelectron-donating ability of the ligand. In addition, the introduction of sulfonate groups into the bis-NHC ligand increased the water solubility of the complex. As a result, a TON of 190,000 was achieved with complex 9 [24]. Interestingly, these complexes also succeeded in the transfer hydrogenation of CO_2 to formate using *i*-PrOH as a hydrogen donor.

3. Decomposition of Fomic Acid

Formic acid is a basic material in the chemical industry. It is used in the production of pharmaceuticals, dyes, perfume ingredients, textiles, and leather. Recently, it has been recognized as a potential *in situ* source of hydrogen for fuel cells because it offers high energy density and can be safely handled in aqueous solutions. However, to date, there is a lack of homogeneous catalysts that are sufficiently active and/or selective for hydrogen production from formic acid.

Compared with the hydrogenation of CO_2 to HCOOH, the back reaction, i.e., the decomposition of HCOOH, is less investigated. However, as mentioned previously, the decomposition of HCOOH is an indispensable step in the hydrogen storage system. The decomposition of formic acid can either liberate CO_2/H_2 by decarboxylation (or dehydrogenation) or give CO/H_2O by decarbonylation (or dehydration) (Equation 1). However, CO acts asa poison for some fuel cell catalysts. Therefore, the efficient production of hydrogen under mild reaction conditions and selective generation of CO-free hydrogen is necessary for practical applications.

$$CO + H_2O \xleftarrow{\text{Decarbonylation}} HCOOH \xrightarrow{\text{Decarboxylation}} CO_2 + H_2 \qquad (1)$$

Ruthenium, rhodium, platinum, and iridium complexes are usually employed for the decomposition reaction. In 1967, Coffey reported pioneering research for the homogeneous decomposition of formic acid in acetic acid under reflux using phosphine-stabilized complexes of a series of noble metals including Pt, Ru, Rh, and Ir [32]. The most active complex, $[IrH_3(PPh_3)_3]$, provided an initial TOF of 8,900 h^{-1} and a TON of more than 11,000. In 1971, Forster used $[Rh(CO)_2Cl]_2$ to catalyze the decomposition of 70% aqueous formic acid at 100 °C in the presence of NaI and obtained a TOF of 4 h^{-1}. However, the corresponding iridium system showed only one third of the above rate under the same conditions [33]. Surprisingly, inlater decades the iridium catalyst was almost ignored. One exception is the research of King et al. who used simple noble metal chlorides (i.e., $RuCl_3$, $RhCl_3$, $PdCl_3$, and $IrCl_3$) for the decomposition of formic acid [34]. The reaction was carried out in a dilute formic acid solution (4.4% in H_2O) under an aerobic condition, which suppressed dehydration. They found that the catalytic activity decreased in the sequence: Ir >> Ru > Pd >> Rh. This study and the ground breaking work of Coffey clearly indicated the high potential of iridium for the decomposition of formic acid. Nevertheless, chemists focused onRu and Rh complexes such as the phosphine complexes (e.g., 14 and 15) and recent N-based complexes (e.g.,16) (Figure 11 and Table 4).

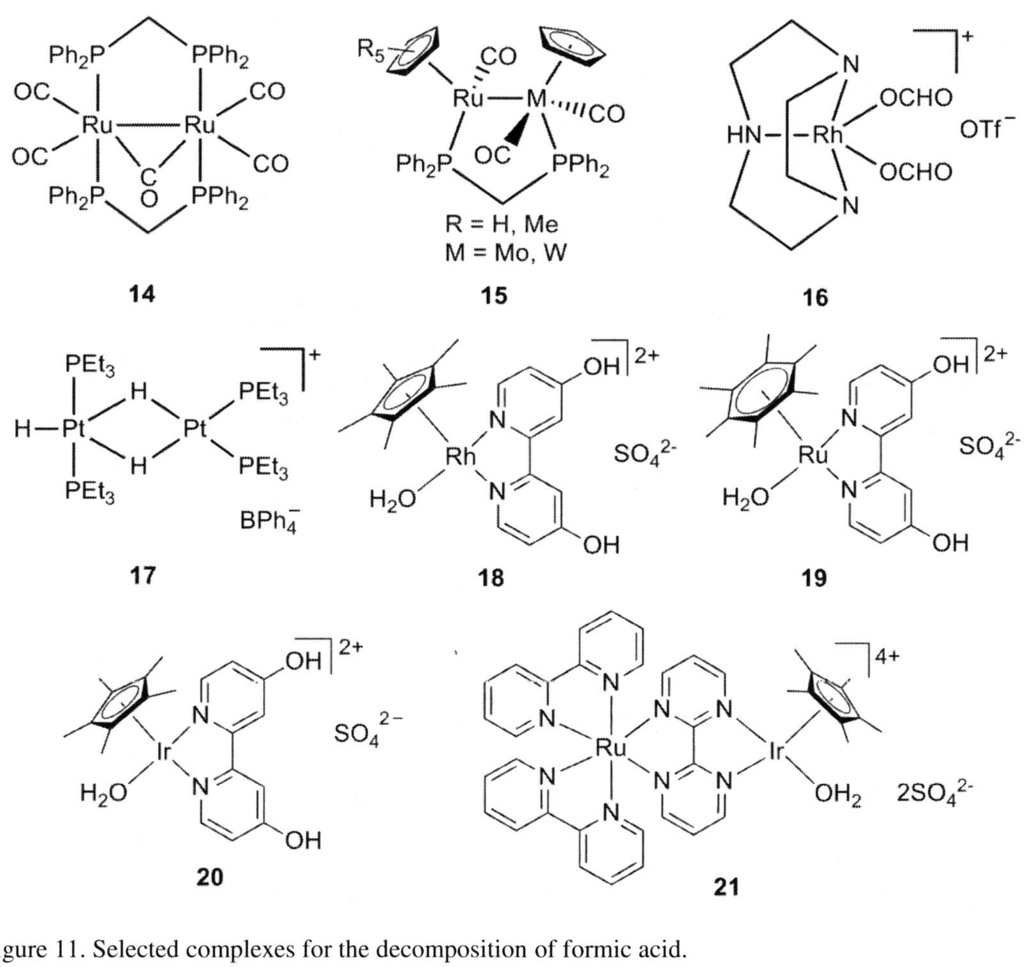

Figure 11. Selected complexes for the decomposition of formic acid.

In 1982, Paonessa and Trogler reported an important study using adinuclear platinum complex, 17. They found that the addition of a small amount of sodium formate accelerated the reaction. This discovery provided a method to improve catalytic efficiency and avoid the use of organic additives. They also found that the initial rate decreased with increased pressure but rose again when the gas was removed. Puddephatt et al. also observed considerably lower activity when the reaction was carried out in a closed tube. They attributed this to presence of CO, which is a strong inhibitor and dissociates from the catalyst $[Ru_2(\mu\text{-}CO)(CO)_4(\mu\text{-}DPPM)_2]$ 14 (DPPM: 1,1-bis(diphenylphosphino)methane) [35].

After decades of effort, chemists have made dramatic progress in the decomposition of formic acid using Rh and Ru complexes. The most effective catalytic systems are listed in Table 4.

The Beller group obtained an unprecedented TON of 260,000 using $[(C_6H_6)RuCl_2]_2$/6DPPE (DPPE: 1,2-bis(diphenylphosphino)ethane) in the presence of an amine additive (i.e., $HexNMe_2$) with the continuous addition of formic acid. [36] Due to the relatively low TOF (900 h^{-1}), the reaction was time consuming and required 264 h for completion. Wills et al. achieved the highest TOF (18,000 h^{-1}) to date using a commercially available complex, $RuCl_2(DMSO)_4$, at 120 °C, although the reaction did require an organic additive (i.e., NEt_3) [37].

Table 4. Decomposition of formic acid using different complexes

Catalyst	Substrate	Solvent	T (°C)	TOF (h^{-1})	TON	ref
$IrH_3(PPh_3)_3$	HCOOH	AcOH	118	8,900	11,000	[32]
16	HCOOH	Me_2CO	r.t.	500	-	[35]
$RuBr_3/3PPh_3$	HCOOH/ NEt_3	-	40	3,630	1,475	[38]
$[(C_6H_6)RuCl_2]_2$ /6DPPE	HCOOH/ $HexNMe_2$	-	r.t.	>900	260,000	[36]
$RuCl_3$/TPPTS	HCOOH/ HCOONa	H_2O	120	460	40,000	[39]
$RuCl_2(DMSO)_4$	HCOOH/ TEA	-	120	18,000	25,000	[37]
[Cp*Rh(bpy)Cl]Cl	HCOOH/HCOONa	H_2O	40	240	-	[27]
18	HCOOH/HCOONa	H_2O	80	7,700	83,000	[29]
20	HCOOH	H_2O	90	14,000	10,000	[40]
	HCOOH	H_2O	40–80	<13,500	100,000	[40]
21	HCOOH/HCOONa	H_2O	r.t.	420	140	[41]

Although Rh and Ru complexes have achieved great success in the decomposition of HCOOH, their status has recently been challenged by Ir complexes. The Himeda group has focused on the application of formic acid as a hydrogen storage material using noble metals, especially iridium [40]. As mentioned above, they found that the DHBP-Ir complex is an efficient catalyst for the hydrogenation of CO_2. More importantly, this complex could also serve as a highly efficient catalyst for the decomposition of HCOOH in water. Most recently, they reported the decomposition of formic acid with DHBP-based rhodium, ruthenium, andiridium aqua complexes [Cp*Rh(DHBP)(H_2O)]SO_4 (18), [C_6Me_6Ru(DHBP)(H_2O)]SO_4 (19), and [Cp*Ir(DHBP)(H_2O)]SO_4 (20) as well as their chloride analogues, [(C_nMe_n)M(4,4'-R_2-2,2'-bpy)Cl]Cl (M = Ir, Rh, n = 5; M = Ru,n = 6; R = OH, OMe, Me, H). [29,40] The reaction was carried out in degassed aqueous formic acid or a formic acid/sodium formate solution at 60–90 °C under an argon atmosphere. Note that no amine additives or organic solvents were required. The catalytic results are presented in Table 5. It is clear that the iridium complex showed higher catalytic activity than the rhodium and ruthenium analogues. The highest TOF (14,000 h^{-1}) was achieved using Ir complex 20 at 90 °C in a 2 M aqueous solution of formic acid (Table5, entry 9).

The study of pH-dependence indicated that the tendency of Ir complex 20 is different thanthat of the Rh and Ru analogues (18 and 19) (Figure 12). Using 20, the maximum catalytic activity was obtained in a 1 M HCOOH (pH = 1.8) solution. The increase of pH by the addition of formate led to decreased initial TOFs. In contrast, using 18 and 19, maximum TOFs were obtained at pH 2.5 and 4.0, respectively. Complete consumption of formic acid was observed in a 1 M HCOOH (pH = 1.8) solution in all cases. However, the reaction was significantly inhibited at higher pH (pH>5) and negligible gas evolution was detected in the sodium formate solution (pH = 7.6).

The study of the electronic substituent effect indicated that the initial TOF values of iridium and ruthenium complexes correlated well with the Hammett constants of the substituents on the ligands (Figure 13). The TOF of 20 with a hydroxyl group ($\sigma_p^+ = -0.91$) was about 90 times that of the unsubstituted analogue, [Cp*Ir(H_2O)bpy]SO_4. In contrast, the TOF of [(C_6Me_6)Ru(4,4'-OH-2,2'-bpy)Cl]Cl was approximately 2.9 times that of the corresponding unsubstituted catalyst. In addition, the carboxyl-substituted Ru complex showed poor catalytic activity due to the electron-withdrawing effect of the carboxyl group ($\sigma_p^+ = 0.42$).

Table 5. The decomposition of formic acid with DHBP-based aquacomplexes[a]

Entry	Catalyst (conc. /mM)	pH	*t* (h)	T (°C)	TOF (h^{-1}) [d]	TON
1	18 (0.4)	1.8	24	60	440	5,000
2	18 (0.2)	2.5	7	60	1,340	4,600
3	18 (0.1)	2.5	2.5	80	7,900	9,300
4 [b]	18	-	50	80	7,700	83,000
5	19 (1.0)	1.8	48	60	36	5,000
6	19 (0.5)	4.0	9	60	140	500
7	19 (0.5)	3.0	5	80	720	1,600
8	20 (0.2)	1.8	5	60	2,400	5,000
9 [c]	20 (0.5)	1.8	2	90	14,000	10,000

[a] The reaction was carried out in a degassed 1 M formic acid/formate solution (10 mL) until gas evolution ceased.

[b] The reaction was carried out in a 4 M formic acid (50 mL)/formate (5 mL) solution using 18 (2 μmol).

[c] The reaction was carried out with in a degassed 2 M formic acid solution.

[d] Initial TOF.

The temperature dependence of the reaction was also investigated through Arrhenius plots (Figure 14). When 20, 18, and 19 were used, the activation energies were calculated to be 76, 93, and 96 kJ mol^{-1}, respectively. Apparently, Ir complex 20 shows higher activationability than the Rh (18) and Ru (19) complexes. Using these catalysts, pressurized hydrogen was successfully generated, which is essential forpractical applications. A spontaneous increase of the gas pressure was observed when the reaction was carried out in a closed system.

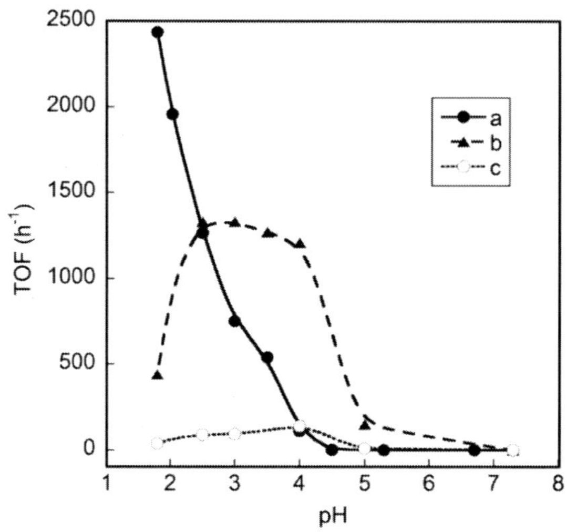

Figure12. pH-Dependence of the reaction rate using a) 20(0.2mM), b) 18 (0.2–0.4 mM), c) 19 (0.5–1.0 mM) at 60 °C in a 1 MHCOOH/HCOONa solution (10 mL).

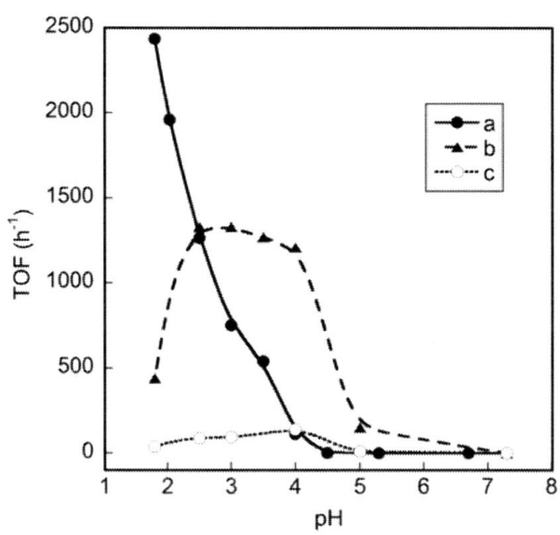

Figure 13. Hammett plot of the initial TOF vs. σ_p^+ value of the substituent (R) for two series of complexes: a) [Cp*Ir(4,4'-R_2-2,2'-bpy)(H_2O)]SO_4, and b) [(C_6Me_6)Ru(4,4'-R_2-2,2'-bpy)Cl]Cl (R = OH, OMe, Me, H, COOH). The reaction was carried out in the presence of catalysts (0.5–2.0 mM) at 60 °C in 10 mL of either 1 M aqueous HCOOH (for Ir complexes) or HCOOH/HCOONa solution (for Ru complexes).

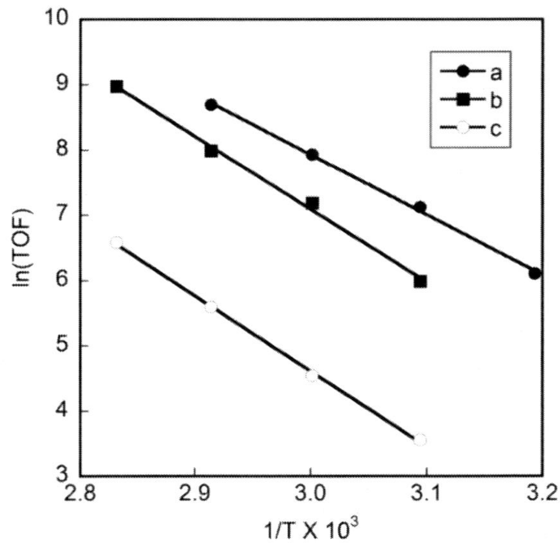

Figure 14. Arrhenius plots of initial TOFs for the decomposition of formic acid using a) 20in a 2 M aqueous HCOOH solution, b) 18 in a 1 M aqueous HCOOH/HCOONa solution (pH = 2.5), and c) 19 in a 1 M aqueous HCOOH/HCOONa solution (pH = 3.0).

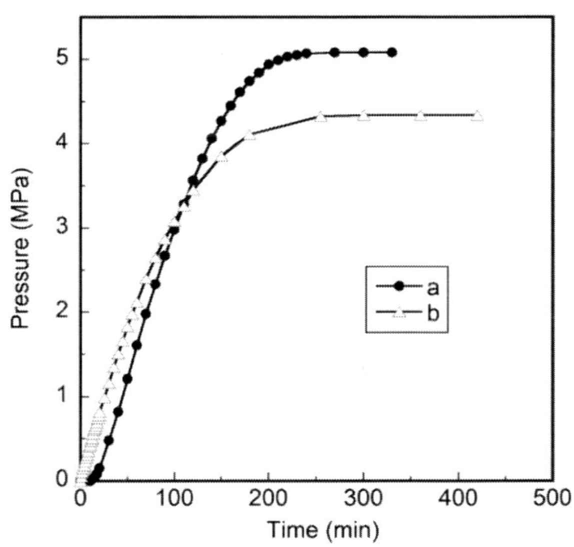

Figure15. a) Time course of reaction pressure using 18 (0.2mM) at 80°Cin a 2 M HCOOH/HCOONa (95:5) solution (10 mL) in a glass autoclave. b) Time course of reaction pressure using 20 (1 μmol) at 80 °C in a 2 M formic acid solution (10 mL) in an autoclave.

As shown in Figure 15, the pressure reached 4–5 MPa using18 and 20. They found no inhibition in this catalytic system since the conversion of formic acid was more than 99%. The salient characteristics of Ir complex 20 are as follows: (1) High catalytic activity (TOF of up to 14,000 h^{-1}) and high thermal stability (90 °C) in aqueous solution. (2) Highly selective transformation of formic acid to H_2 and CO_2. It can also provide CO-free hydrogen by controlling the temperature. (3) Complete conversion offormic acid. (4) Ability to produce high-pressure hydrogen.

Based on theexperimental studies, they proposed the mechanism shown in Figure 16. The replacement of aqua ligand in 20 by formic acid leads to aformato complex. β-Hydrogen elimination generates ahydride intermediate and results in CO_2 evolution. The hydride then reacts with hydroniumto deliver hydrogen and regenerate complex 20.

In 2010, Fukuzumi group reported a heterodinuclear iridium-ruthenium catalyst, $[Ir^{III}(Cp^*)(H_2O)(bpym)Ru^{II}(bpy)_2](SO_4)_2$ (bpym: bipyrimidine) (21), for the decomposition of formic acid in water [41]. The catalytic activity of this complex was pH dependent and showed the highest TOF (426 h^{-1}) at ambient temperature at pH 3.8, which is in agreement with the pKa value (3.75) of HCOOH. The dependence of TOF on the temperature was also studied. The activation energy calculated from the Arrhenius plot (71 kJ mol^{-1}) is much lower than that without catalyst (326 kJ mol^{-1}).

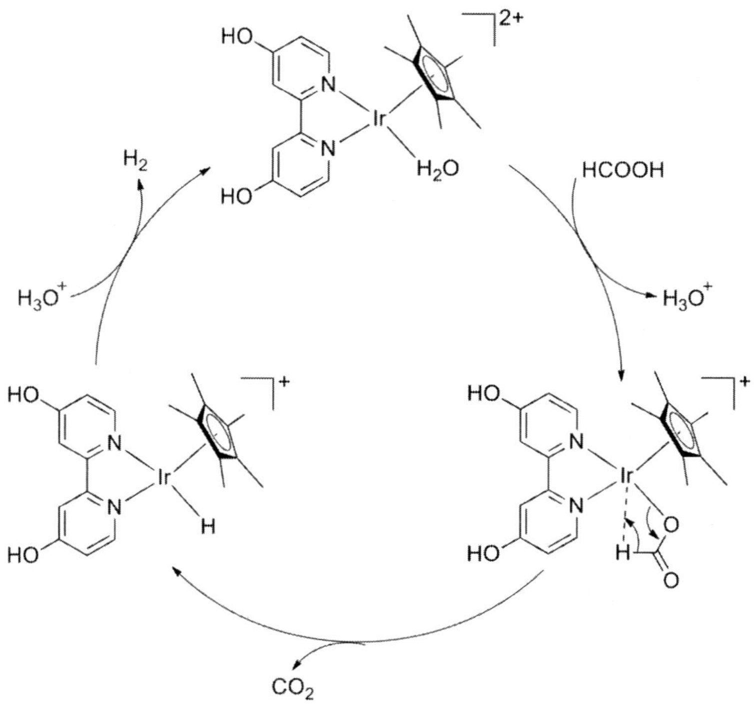

Figure 16. Proposed mechanism for the decomposition of formic acid.

4. WATER OXIDATION

Water oxidation is another central reaction in natural photosynthesis. It is known as a half reaction of the water-splitting redox reaction, which is considered to be a sustainable method for H_2 production from H_2O using artificial photosynthesis.

However, in comparison to the water reduction process, water oxidation is more complex, more thermodynamically demanding, and kinetically slower since it requires the loss of four electrons and four protons from two water molecules and the formation of an oxygen-oxygen bond [42]. Thus high efficiency water oxidation is the key to solving the water-splitting problem. Water oxidation requires a minimum electrical potential of 1.23 V at pH = 0 (Equation 2).

Thermodynamically, Ce^{IV} is capable of oxidizing water owing to its redox potential (E^0 = 1.67 V) but it requires a catalyst to lower the large activation barriers (Equation 3). Recently, the development of efficient water oxidation catalysts has attracted considerable attention [43−45].

$$2H_2O \longrightarrow O_2 + 4H^+ + 4e^- \qquad E^0 = 1.23 \text{ V vs. NHE (pH = 0)} \tag{2}$$

$$4Ce^{4+} + 2H_2O \xrightarrow{\text{Catalyst}} 4Ce^{3+} + 4O_2 + 4H^+ \tag{3}$$

In 1982, Meyer and co-workers reported thefirst homogeneous water-oxidation catalyst, the so-called "blue dimer" $[(Ru(bpy)_2(OH_2))_2(\mu\text{-}O)]^{4+}$ 22 (Figure 17). [46,47] The blue dimer oxidized water with a TON of 13.2 using an excess of primary oxidants such as cerium(IV) ammonium nitrate (CAN); however, it lost its catalytic activity after just a few cycles.

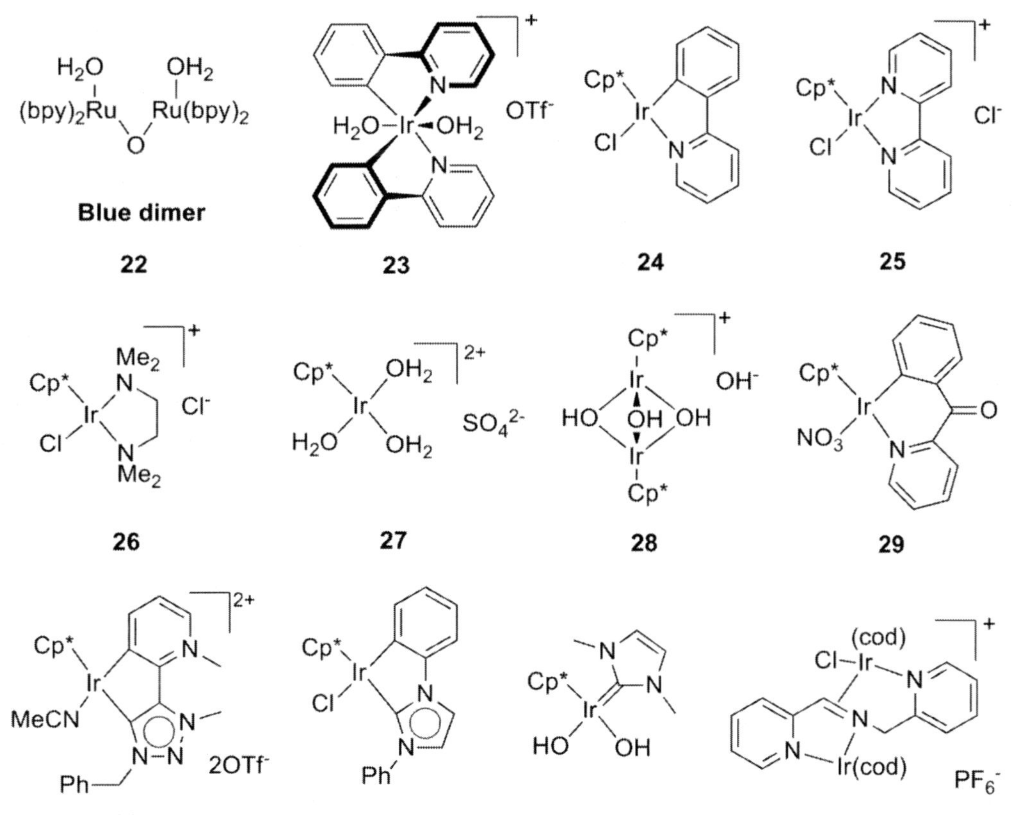

Figure 17. Typical iridium complexes for water oxidation.

Since then, many catalytic variations have been reported based on binuclear ruthenium and polypyridyl ligands. However, the investigation of iridium complexes was triggered only three years ago by the landmark research of the Bernhard group. In 2008, Bernhard and co-workers reported the first mono-nuclear iridium catalyst 23, $[Ir(H_2O)_2(ppy)_2]^+$, which oxidized water in the presence of CAN [48]. The catalyst showed high a TON of 2,500 with an efficiency of 66% with regard to CAN; however, the TOF was much lower than that of the ruthenium complexes. The homogenous bis-phenylpyridine (bis-ppy) bis-aquoiridium(III) complex was robust in an aqueous medium due to the strong carbon-iridium bond. Moreover, they found that the energies of the HOMOs were tunable by alteringthe substitution pattern of the cyclo-metalated ligands. Thus, the electronics of the complexes could be adjusted to optimizethe reaction.

Crabtree and co-workers predicted that an electron-rich environment could stabilize high-valent $Ir^V{=}O$ or Ir^{IV}-O· intermediates for water oxidation. Thus, they employed a very strongly donating ligand toconstructa series of cyclopentadienyliridium catalysts (e.g., 24–28) with polypyridine, phenylpyridine, and diamines, etc. [49,54,55] The half-sandwich iridium complexes with phenylpyridine (24) and bipyridine (25) showed relatively highactivity for the homogeneous oxidation of water to dioxygen with CAN as the primary oxidant (Table 6).

Kinetic studies on 26 indicated that *N,N,N',N'*-tetramethylenediamine (tmeda) is rapidly protonated and liberated from the iridium center to generate the same precursor, $[Cp^*Ir(H_2O)_3]^{2+}$, as 27. Accordingly, 26 and 27 showed similar catalytic activity under the same reaction conditions. Among these iridium complexes, dinuclear complex 28 showed the highest catalytic activity and gave a TOF of 1,200 h^{-1}. Ruthenium and rhodium analogues, i.e., Cp*RuCl(bpy) and [Cp*RhCl(bpy)]Cl, were also investigated but no oxygen evolution was observed. The key intermediate for water oxidation was believed to be an $Ir^V{=}O$ complex, as shown in Figure 18. The proposed mechanism involving sequential one-electron oxidation of the complex via proton-coupled electron transfer was supported by DFT calculations. Inspired by Crabtree's studies, a number of half-sandwich iridium catalysts were reported.

Complex 24 does not fully dissolve in water until CAN isadded. To improve the water solubility of complex, Macchioni and co-workers designed and synthesized a robust and water-soluble iridium catalyst, 29, using NO_3^- instead of Cl^-. Additionally, they introduced abridging -C(=O)- electron-withdrawing group between the two aryl rings [50].

Table 6. Catalytic activities of iridium catalysts for water oxidation

Catalyst	Time (h)	TOF (h^{-1})	TON	ref
22	-	15	-	[46]
23	168	18	2,500	[48]
24	-	600	-	[49]
25	-	860	-	[49]
26	-	370	-	[49]
27	-	330	-	[49]
28	-	1,200	-	[49]
29	-	510	-	[50]
30	120	314	10,000	[51]
32	1	5,400	>2,000	[52]
33	2	3,400	>1,000	[53]

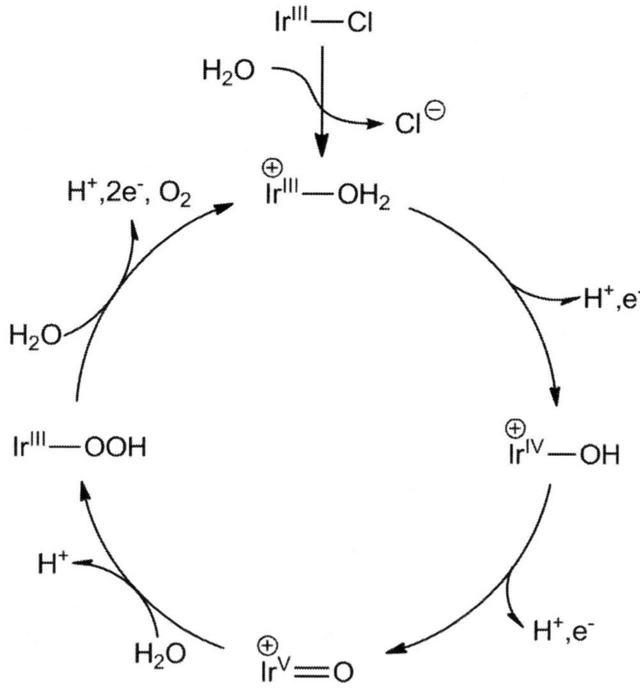

Figure 18. Proposed mechanism for water oxidation by Crabtree et al.

When 29 was used for the water oxidation reaction, CAN was quantitatively consumed. The TONs ranged from 50 to 1,250 and a TOF of 510 h^{-1} was obtained. More importantly, 29 showed longterm activity that was about 2 to 3 times higher than that of 24. Their study indicated that the degradation of 24 and 29 during the reaction does not result in a tris-aqua complex, $[Cp^*Ir(H_2O)_3](NO_3)_2$, because this complex showed a much higher TOF (940 h^{-1}) than 24 and 29. Based on the rationale that abnormal carbines with zwitterionic resonance forms may stabilize different metal oxidation states, Albrecht and co-workers investigated iridium complexes with two different abnormally bound *N*-heterocyclic carbene ligands. [51] Complex 30 gave the highest TON of 10,000, which was obtained after five days, although the reaction rate is relatively low (TOF of 314 h^{-1} after 10 h). They attributed the high activity to the electronic flexibility of the mesoionic ligands, which stabilize the high-oxidation-state Ir^V oxo species. Based on a similar consideration, Crabtree et al. also investigated an NHC-iridium complex, 31. Using electron paramagnetic resonance (EPR) spectroscopy, they demonstrated that the high-valent intermediate, $[Cp^*Ir^{IV}(NHC)Cl]^+$, of 31 is much more stable than that of 24. Moreover, they found that the ligands do not dissociate from the metal center to from fragments such as $[Cp^*Ir(H_2O)_3]^{2+}$ viacyclic voltammetry studies. The combined research of these groups suggests that the ligand plays an important part in the catalytic process. The cooperative action between the transition metal and ligand should be noted for further complex design.

Hetterscheid et al. used *N*-dimethylimidazolin-2-ylidene as a ligand to construct another NHC-based iridium complex [52]. The half-sandwich iridium catalyst 32 showed the highest TOF (5,400 h^{-1}) to date and reached a TON of 2,000 (Table 6). Using electro spray ionization mass spectrometry (ESI-MS), they observed a set of peaks corresponding to $[Cp^*Ir{=}O]^+$ and $[Cp^*Ir(OH)_3]^+$ in the reaction mixture. This indicates that 32 was actually oxidized and the reaction proceeded through a similar mechanism as other mononuclear catalysts. De Bruin and co-workers synthesized a cationic binuclear iridium compound, 33, which exhibited a high TOF of 3,400 h^{-1} and a TON of >1,000 after 2 h (Table 6) [53]. However, the mononuclear iridium complex, [(cod)Ir(Cl)(ppei)] (ppei: 2-pyridinal-1-phenylethylimine), showed a substantially lower activity (TOF of 370 h^{-1} and TON of >100). The synergistic effects of the two metal centers may contribute to the high activity.

Most recently, Crabtree et al. reported electrochemical water oxidation using 27 and 28 [56]. The deposition of 27 and 28by electrochemical oxidation formed a blue layer of an iridium oxide material (BL), which could catalyze water oxidation at the anode. The TOF was estimated to be 5,200 h^{-1} at 1.4 V for 4.1 nmol cm^{-2} BL. The high activity of BL was attributed to its high surface area since SEM analysis showed nanoscale features. They suggested that inorganic iridium dioxide is the active catalyst in this case because of the striking similarity in electrocatalytic behavior andvoltammetric response of BL and other iridium oxides.

5. Transfer Hydrogenation

The transfer hydrogenation (TH) reaction using formate as a hydrogen donor in water has received much attention in green chemistry because of its simple operation and avoidance of the use of explosive molecular hydrogen and harmful organic solvents. TH has emerged as a powerful and practical tool for organic transformations in both academia and industry. Ruthenium and rhodium catalysts play a major role in these fields [57,58]. Nevertheless, iridium complexes have recently been applied. The homogenous TH of C=O and C=N catalyzed by Ir complexes in water using formate as a donor will be reviewed here. In 1990, Joo used a TPPMS-based iridium catalyst to catalyze the biphasic TH of benzaldehyde; however, the reactiononly gave 11% conversion while complete conversion was achieved using the ruthenium analogue, RuCl $(TPPMS)_2$ [59].

Compared to the phosphine-based Ir complex, the amine-based Ir complex showed better catalytic performance especially with a Cp* ligand incorporated. Since the pioneering study on the reduction of NAD^+ to NADH (NADH: nicotinamide adenine dinucleotide) in water catalyzed by a homogenous pentamethylcyclopentadienyl rhodium complex, $[Cp^*Rh(bpy)Cl]^+$, was reported by Steckhan, [60,61] considerable efforts have been devoted to the iridium analogues (Cp*Ir) due to their water-solubility, high stability, and effectiveness. Ogo et al. found that an Ir complex, i.e., $[Cp^*Ir(H_2O)_3]^{2+}$, showed pH-dependent catalytic activity for the TH of aldehyde in water using formate as a donor [62,63]. Although the maximum TOF at pH 3.2 was very poor (<4 h^{-1}), the potential of iridium catalystsfor TH in water was revealed. Using bipyridine as a second ligand led to improved catalytic activity. Himeda reported that the use of $[Cp^*Ir(bpy)Cl]^+$ for theTH of acetone at pH 3.5 showed a TOF of 41 h^{-1}, which was similar to that of the rhodium analogue [28]. Furthermore, aqua Ir complex 20, which contains a proton-responsive DHBP ligand, showed highly pH-dependent catalytic activity and chemoselectivityfor aqueous TH (Figure 19) [64]. The transfer hydrogenation of 2-cyclohexen-1-one gave cyclohexanone with a high TOF of 4,300 h^{-1} under basic conditions, while cyclohexanol gave amoderate TOF of 1,100 h^{-1} under acidic conditions; selected examples are listed in Table7.

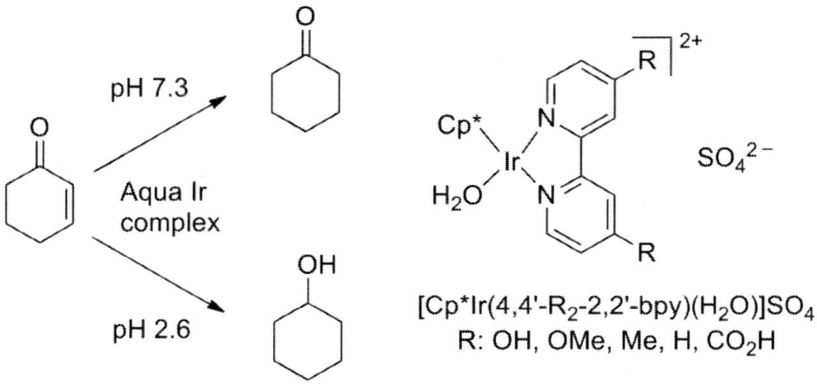

Figure 19. Transfer hydrogenation of 2-cyclohexen-1-oneusingIr complexes.

Table 7. pH-Dependenttransfer hydrogenation catalyzed by 20[a]

Entry	Substrate	pH2.6			pH7.3		
		TOF (h^{-1})	Yield (%)	Product	TOF (h^{-1})	Yield (%)	Product
1	cyclohexen-2-one	1,100	94	cyclohexanol	4,300	>99	cyclo-hexanone
2	1-butene-2-one	1500	72	2-butanol	5,300	>99	2-butanone
3	2-hexene-4-one	700	29	3-hexanol	4,100	>99	3-hexanone
4	cinnamaldehyde	1,200	85	cinnamyl alcohol	4,100	17	hydro-cinnam-aldehyde

[a] The reaction was carried out at 40 °C in an aqueous 1 M formate solution (20 mL) with a substrate (2 mmol) and catalyst 20 (0.5 μmol); formate/substrate/catalyst = 40000/4000/1.

The TOF profiles for TH using DHBPIrcomplex 20 and the methoxy analogue, ([Cp*Ir(4,4'-$(MeO)_2$-2,2'-bpy)(H_2O)]SO_4), were markedly different. The substantial change in activity was due to the electronic effect based on the above-mentioned acid–base equilibrium of the phenolic hydroxyl group of DHBP (Figure 20).

Figure 21 shows Hammett plots to portray the electronic substituent effects on the TH of2-cyclohexen-1-one using a series of iridium complexes, i.e., [Cp*Ir(4,4'-R_2-2,2'-bpy)(H_2O)]SO_4 (R: OH, OMe, Me, H, CO_2H) at pH 2.6 and 7.3. Note that the phenolic hydroxyl (σ_p^+ = –0.92) and carboxyl (σ_p^+ =+0.42) groups were deprotonated to generate the oxyanion (σ_p^+ =–2.30) and carboxylate (σ_p^+ =–0.02) under basic reaction conditions. [65] Under these conditions, the TOF of the DHBP complex was more than 1,000 times that of the unsubstituted analogue; the highest TOF (81,000 h^{-1}) was obtained for the TH of benzaldehyde at pH 7.3 and 80 °C. From theplot of the absorption spectra of 20 as a function of pH, the acid–base equilibrium of the phenolic hydroxyl groups and the aqua ligands were examined. The change in the optical density at 265 nm plotted against the pH is shown in Figure 22.

Thus, the complex appeared to exist in a hydroxyl form at pH 2.6, an oxyanion form at pH 7.3, and as a hydroxo complex at pH values above 10.

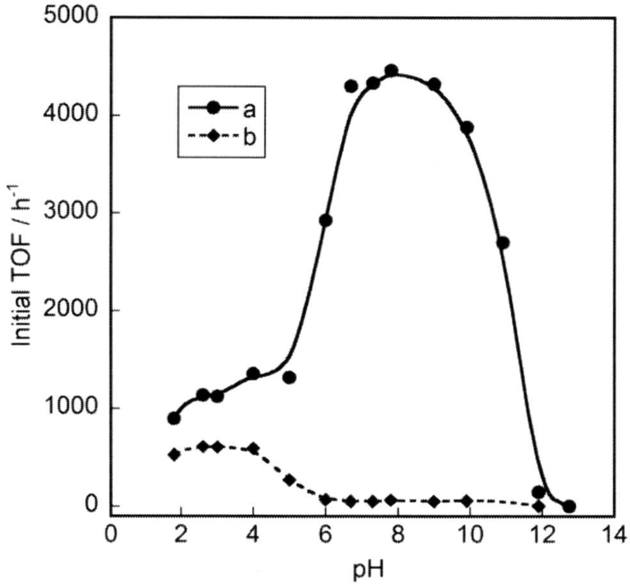

Figure 20. pH-Dependent profile of the initial TOFs for the TH of 2-cyclo-hexen-1-one (2.0 mmol) witha) 20 (0.5 μmol) and b) [Cp*Ir(4,4'-(MeO)$_2$-2,2'-bpy)(H_2O)]SO_4 (2.0 μmol) in an aqueous 1M formate solution (20 mL) at 40 °C.

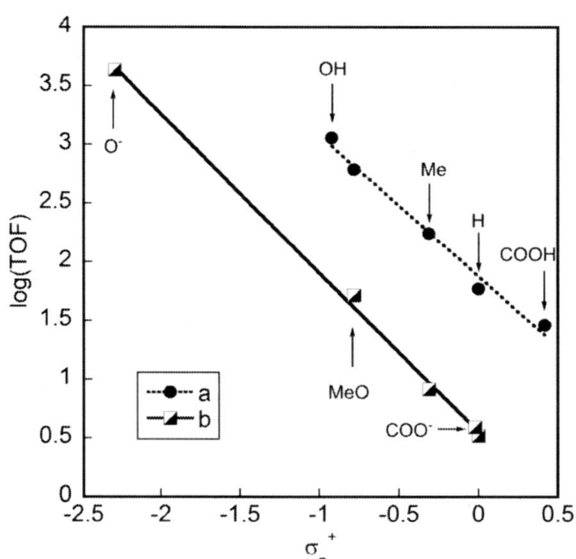

Figure 21. Plot of the correlation between σ_p^+ and the initial TOF for the TH of 2-cyclo-hexen-1-one at a) pH 2.6 and b) pH 7.3 with catalyst [Cp*Ir(4,4'-R$_2$-2,2'-bpy)(H_2O)]SO_4 (R: OH, OMe, Me, H, CO_2H) (0.5–20 μmol) at 40 °C.

Interestingly, the TOF ratio of DHBP-Ir complex 20 and the methoxy analogue, [Cp*Ir(4,4'-(MeO)$_2$-2,2'-bpy)(H_2O)]SO_4, which features a pH-independent electronic effect, overlapped appreciably with the curve of absorbance change. This result strongly suggests that the increase in activity can be attributed to the increase in the electron-donating effect due to the deprotonation of the hydroxyl group.

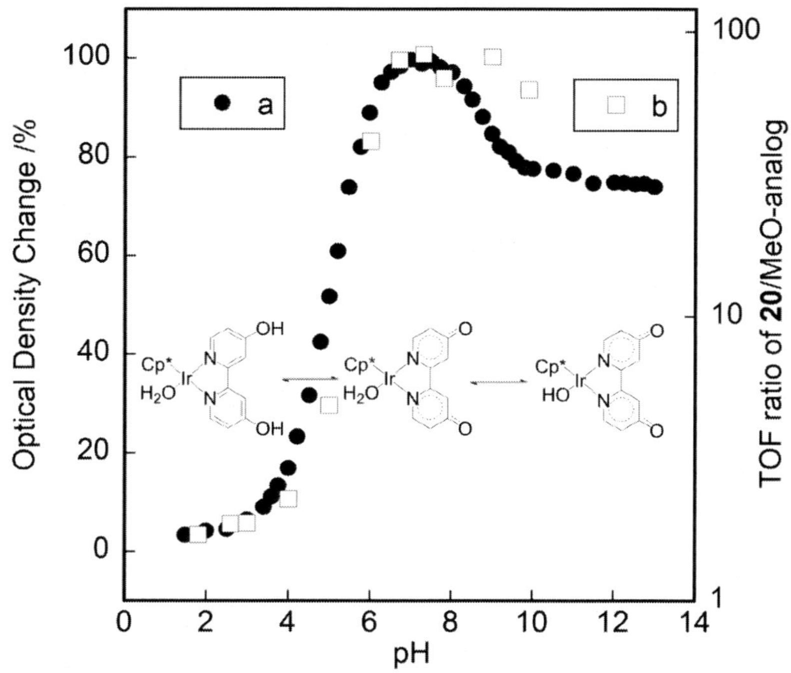

Figure 22. a) Absorbance change of 20 at 265 nm as a function of pH.b) TOF ratio of 20 vs. [Cp*Ir(4,4'-$(MeO)_2$-2,2'-bpy)(H_2O)]SO_4 as a function of pH on a logarithmic scale.

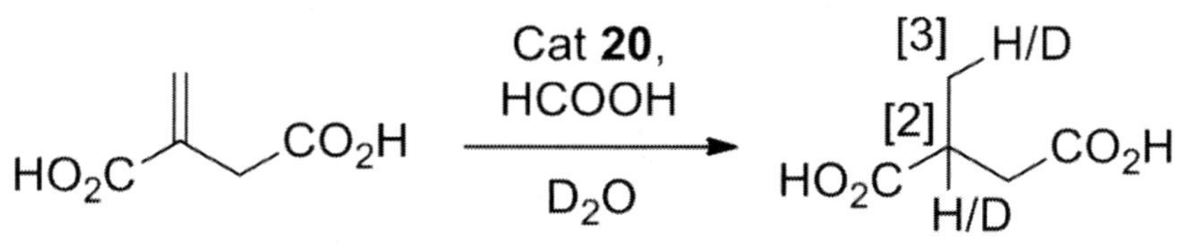

Figure 23. Transfer deuterogenation catalyzed by Ir complex 20.

Further studies on TH in D_2O provided a cost-effective and easy method for the preparation of regioselective deuterium-labelled compounds and useful mechanistic knowledge (Figure 23) [66].

The TH of itaconic acid in D_2O using HCOOH gave 2,3-dideuterated methylsuccinic acid in a quantitative yield. The transfer deuterogenation was applied to the preparation of α-deuterated alcohols from ketones in high yields and with good to high deuteron content (Table 8). Positive-ion ESI-MS analysis showed that the reaction of 20 with H_2 in D_2O at a pD of 2.7 afforded the *O*-deuterated deuteride $[Cp^*IrD(D_2L)]^+$. Interestingly, in the TH of itaconic acid using DCO_2D in H_2O, methylsuccininc acid with low D content was produced.

Based on the experimental results, the authors proposed a mechanism for deuterogenation catalyzed by 20 ([Ir]–X) using HCOOH in D_2O, as shown in Figure 24.

Table 8. Transfer deuterogenationof ketones to α-deuterated alcoholsusing 20[a]

Entry	Substrate	Time (h)	Catalyst loading (μmol)	Yield (%)	D content (%) [b]
1	Cyclopentanone	36	2.5	98	92
2	Cyclohexanone	20	0.5	>99	74
3	Cycloheptanone	20	1.0	>99	85
4	Acetophenone	24	2.0	>99	73
5	Propiophenone	24	2.0	>99	80
6	2-Cl(C_6H_4)COMe	48	2.0	99	87

[a] The reaction was carried out using HCOOH in D_2O at 50 °C.
[b] Determined by ^1H NMR.

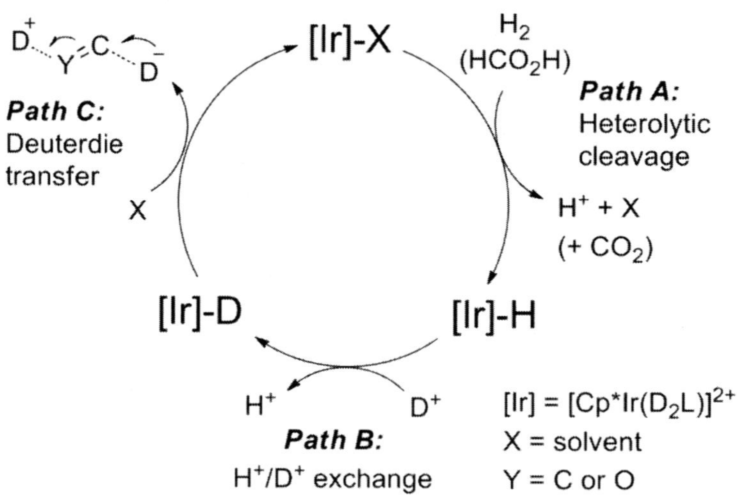

Figure 24. Proposed mechanism for deuterogenation.

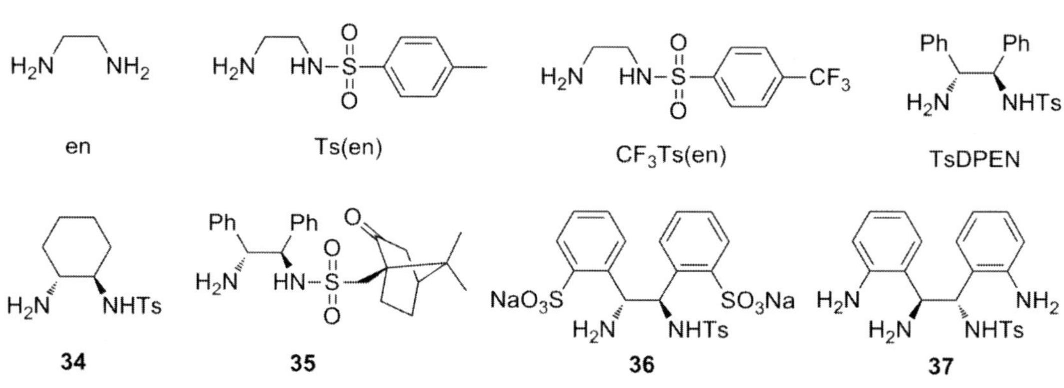

Figure 25. Selected diamine ligands for TH.

Iridium hydride ([Ir]–H) is readily generated by the reaction of 20 via the decomposition of HCOOH (Path A). [Ir]–H immediately exchanges with a deuteron of D_2O to produce iridium deuteride ([Ir]–D) and a proton (Path B).

The deuteride of [Ir]–D is transferred to the carbonyl carbon of a ketone or the β-carbon of an α, β-unsaturated carboxylic acid (Path C). During external deuterogenation from D_2O via ionic hydrogenation or a monohydride mechanism, dideuterated compounds are obtained. Conceptually, this is the first approach for the production ofmetal deuteride derived from the deuteron of D_2O with a high reducing ability. In the asymmetric transfer hydrogenation (ATH) field, ruthenium and rhodium catalysts with chiral phosphine ligand are predominant until 2000. Inspired by the landmark work of Noyori on mono to sylated diamines and 1,2-aminoalcohols for the ATH of aromatic ketones [67,68], the Xiao group recently performed extensive studies on ATH using iridium with diamines and amino alcohols ligands in addition to the rhodium and ruthenium analogues [57,58].

Table 9. TH of benzaldehyde in water at 65 °C

Entry	Metalprecursor	Ligand	H donor	TOF (h^{-1})
1	$[RuCl_2(p\text{-cymene})]_2$	en	HCOONa	12
2	$[Cp^*RhCl_2]_2$	en	HCOONa	900
3	$[Cp^*IrCl_2]_2$	en	HCOONa	1,800
4	$[RuCl_2(p\text{-cymene})]_2$	Ts(en)	HCOONa	1,000
5	$[Cp^*RhCl_2]_2$	Ts(en)	HCOONa	6,000
6	$[Cp^*IrCl_2]_2$	Ts(en)	HCOONa	12,000
7	$[Cp^*IrCl_2]_2$	Ts(en)	*i*-PrOH	26
8	$[Cp^*IrCl_2]_2$	Ts(en)	$HCOO/Et_3N$	15
9 [a]	$[Cp^*IrCl_2]_2$	$CF_3Ts(en)$	HCOONa	132,000

[a] Reaction was carried out at 80 °C.

In 2006, they reported a highly efficient and chemoselective TH of a wide range of aldehydes using Ir precursors and *N*-tosylethylenediamine ligands (Figure 25) in water and air. [69] The effects of the central metals, ligands, and hydrogen donors were examined (Table 9). The introduction of an electron-with drawing CF_3 group at the tosylate end of the diamine led to a remarkable increase of the TOF (132,000 h^{-1}) in the TH of benzaldehyde at 80°C
(Entry 9, Table 9).

They also reported asymmetric transfer hydrogenation (ATH) using Cp*Ir and modified chiral diamine ligands (i.e., 34–37) as efficient catalysts fora wide range of ketone substrates with high yields and ee values using formate as a hydrogen donor in water and air [70–73].

Although the Ir complex was less active than the Rh and Ru analogues in most cases, Cp*Ir-35 out performed the Ru and Rh analogues in both catalytic activity and enantioseletivity. For example, using Cp*Ir-35, acetophenone was reduced with 97% conversion and 98 %ee in 2.5 h at a substrate/catalyst ratio of 1,000.

Moreover, Cp*Ir- and Cp*Rh-TsDPEN catalysts showed a similar pH-dependency of the reaction rate as the $[Cp^*Ir(bpy)]^+$ complex. The highest rate was obtained at pH 7.0–7.5. Thus, under neutral conditions they investigatedthe ATH of various ketones in aqueous HCOONa; selected results are presented in Table 10. Almost full conversion and excellent enantioselectivity were achieved. In addition, they proposed a mechanism to illustrate the pH-dependent activity and enantioselectivity (Figure 26). Under neutral conditions, the ATH proceeds via an efficient catalytic cycle, I, that is similar to the procedure postulated by Noyori [74,75]. However, in acidic solutions, the reduction occurs through cycle II, which is less effective and stereo selective. When the solution is highly basic, the catalyst will transform to a hydroxo species, which does not participate in the catalytic cycle. Thus, the reaction at pH 9–11 is inhibited due to the lowered concentration of active species but maintains its enantioselectivity.

Table 10. ATH of ketones with Rh- and Ir-(*R*,*R*)-TsDPEN in water[a]

Entry	Ketone	Rh cat.		Ir cat.	
		Conv.[b]	ee[b]	Conv.[b]	ee[b]
1	PhCOMe	99	97	99	93
2	4-ClC$_6$H$_4$COMe	>99	94	>99	82
3	4-NO$_2$C$_6$H$_4$COMe	98	88	98	73
4	4-MeC$_6$H$_4$COMe	98	93	96	88
5	3-MeOC$_6$H$_4$COMe	97	98	97	90
6	4-MeOC$_6$H$_4$COMe	90	97	97	91
7	α-tetralone	96	99	96	97
8	3-acetylpyridine	99	98	>99	60

[a] Reaction conditions: 1 mmol of ketone, 3–8 eq of HCOONa, S/C = 100, in 2 mL of water, 40 °C.
[b] Determined by GC. The configurations of the alcohols are *R*.

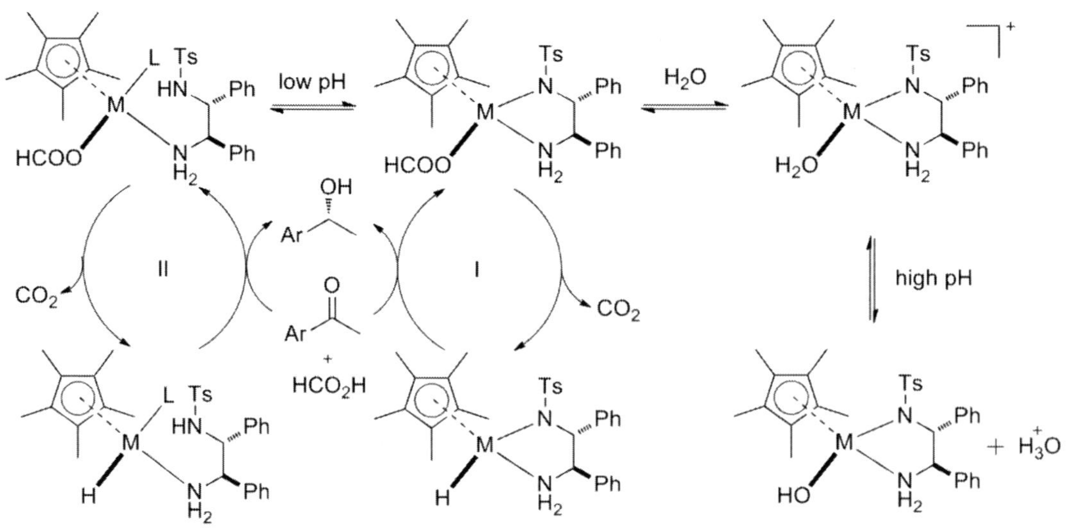

Figure 26. Proposed mechanism for the ATH of ketones in water with formate as a donor.

Deng et al. reported water-soluble disulfonated *N*-tosylethylenediamine 36 combined with Rh, Ru, and Irfor the ATH of ketones [76] and cyclic imines and iminiums [77] in water using formate. However, the iridium complex showed lower conversion and enantioselectivity than the ruthenium and rhodium catalysts. In 2007, they reported another water-soluble ligand, 37, for the ATH of ketones and imines in aqueous media [78]. Once more, the Ir catalyst proved to be less effective than the Rh and Ru analogues.

CONCLUSION

Extensive research has recently discovered the excellent performance of iridium complexes. They have been proven to be exceptionally versatile catalysts for bio-inspired reactions, which have the potential to contribute to energy and environmental chemistry. Complexes of Cp*Ir combined with tailor-made ligands have shown high catalytic activity for formic acid formation and decomposition. Moreover, the homogenous Ir complex has fulfilled the requirements to be recyclable and reusable for CO_2 hydrogenation in aqueous solutions. These results convincingly demonstrate the feasibility of formic acid as a hydrogen storage material. Using iridium complexes, high catalytic activity and stereoselectivity have also been achieved in the asymmetric variation of transfer hydrogenation, which is a safe and valid alternative to the traditional hydrogenation reaction. For water oxidation, iridium complexes show much higher catalytic activity and durability than the usually used ruthenium complexes.

Although great success has been achieved, these catalytic reactions are still far from practical application. There are many problems that urgently need to be solved. For instance, in the homogeneous catalytic hydrogenation of CO_2, high reaction performance generally requires an elevated temperature and pressure, which decreases the energy efficiency. For water oxidation, the catalytic efficiency, total turnover numbers and catalyst stability require further improvement and the mechanism requires more investigation for full elucidation. From practical and environmentally benign perspectives, substantial work should concentrate on the development of extraordinary catalysts that function under mild reaction conditions in a greener manner. Sophisticated multifunctional ligands and synergistic effects between the metal and ligand should be considered for the design of new catalysts. Since homogenous catalysis using iridium complexes has progressed rapidly, much more exciting results are expected in the near future.

REFERENCES

[1] Knowles, W. S. *Acc. Chem. Res.* 1983, 16, 106.

[2] Noyori, R.; Ohta, M.; Hsiao, Y.; Kitamura, M.; Ohta, T.; Takaya, H. *J. Am. Chem. Soc.* 1986, 108, 7117.

[3] Vaska, L. *Acc. Chem. Res.* 1968, 1, 335.

[4] Crabtree, R. *Acc. Chem. Res.* 1979, 12, 331.

[5] Leitner, W.; Dinjus, E.; Gassner, F. In *Aqueous-Phase Organometallic Catalysis, Concepts and Applications*; Cornils, B., Herrmann, W. A., Eds.; Wiley-VCH: Weinheim, 1998, p 486.

[6] Jessop, P. G.; Ikariya, T.; Noyori, R. *Chem. Rev.* 1995, 95, 259.

[7] Jessop, P. G.; Joó, F.; Tai, C. C. *Coord. Chem. Rev.* 2004, 248, 2425.

[8] Jessop, P. G. In *Handbook of Homogeneous Hydrogenation*; De Vries, J. G., Elsevier, C. J., Eds.; Wiley-VCH: Weinheim, 2007; Vol. 1, p 489.

[9] Himeda, Y. *Eur. J. Inorg. Chem.* 2007, 3927.

[10] Federsel, C.; Jackstell, R.; Beller, M. *Angew. Chem. Int. Ed.* 2010, 49, 6254.

[11] Wang, W.; Wang, S.; Ma, X.; Gong, J. *Chem. Soc. Rev.*2011, 40, 3703.

[12] Enthaler, S.; von Langermann, J.; Schmidt, T. *Energy Environ. Sci.* 2010, 3, 1207.

[13] Enthaler, S. *ChemSusChem* 2008, 1, 801.

[14] Joó, F. *ChemSusChem* 2008, 1, 805.

[15] Inoue, Y.; Izumida, H.; Sasaki, Y.; Hashimoto, H. *Chem. Lett.* 1976, 863.

[16] Joó, F.; Laurenczy, G.; Nadasdi, L.; Elek, J. *Chem. Commun.* 1999, 971.

[17] Erlandsson, M.; Landaeta, V. R.; Gonsalvi, L.; Peruzzini, M.; Phillips, A. D.; Dyson, P. J.; Laurenczy, G. *Eur. J. Inorg. Chem.* 2008, 620.

[18] Horvath, H.; Laurenczy, G.; Katho, A. *J. Organomet. Chem.* 2004, 689, 1036.

[19] Himeda, Y.; Onozawa-Komatsuzaki, N.; Sugihara, H.; Kasuga, K. *Organometallics* 2007, 26, 702.

[20] Tanaka, R.; Yamashita, M.; Nozaki, K. *J. Am. Chem. Soc.* 2009, 131, 14168.

[21] Schmeier, T. J.; Dobereiner, G. E.; Crabtree, R. H.; Hazari, N. *J. Am. Chem. Soc.*2011, 133, 9274.

[22] Sanz, S.; Benitez, M.; Peris, E. *Organometallics* 2010, 29, 275.

[23] Sanz, S.; Azua, A.; Peris, E. *Dalton Trans.* 2010, 39, 6339.

[24] Azua, A.; Sanz, S.; Peris, E. *Chem. - Eur. J.* 2011, 17, 3963.

[25] Himeda, Y.; Onozawa-Komatsuzaki, N.; Sugihara, H.; Kasuga, K. *J. Am. Chem. Soc.* 2005, 127, 13118.

[26] Himeda, Y.; Onozawa-Komatsuzaki, N.; Sugihara, H.; Kasuga, K. *J. Photochem. Photobiol. A* 2006, 182, 306.

[27] Himeda, Y.; Onozawa-Komatsuzaki, N.; Sugihara, H.; Arakawa, H.; Kasuga, K. *Organometallics* 2004, 23, 1480.

[28] Himeda, Y.; Onozawa-Komatsuzaki, N.; Sugihara, H.; Arakawa, H.; Kasuga, K. *J. Mol. Catal. A: Chem.* 2003, 195, 95.

[29] Himeda, Y.; Miyazawa, S.; Hirose, T. *ChemSusChem* 2011, 4, 487.

[30] Ahlquist, M. S. G. *J. Mol. Catal. A: Chem.* 2010, 324, 3.

[31] Yang, X. *ACS Catal.* 2011, 1, 849.

[32] Coffey, R. S. *J. Chem. Soc., Chem. Commun.* 1967, 923a.

[33] Forster, D.; Beck, G. R. *J. Chem. Soc., Chem. Commun.*1971, 1072.

[34] King, R. B.; King, A. D.; Bhattacharyya, N. K. *Transition Met. Chem.* 1995, 20, 321.

[35] Gao, Y.; Kuncheria, J.; Yap, G. P. A.; Puddephatt, R. J. *Chem. Commun.* 1998, 2365.

[36] Boddien, A.; Loges, B.; Junge, H.; Gartner, F.; Noyes, J. R.; Beller, M. *Adv. Synth. Catal.*2009, 351, 2517.

[37] Morris, D. J.; Clarkson, G. J.; Wills, M. *Organometallics* 2009, 28, 4133.

[38] Boddien, A.; Loges, B.; Junge, H.; Beller, M. *ChemSuschem* 2008, 1, 751.

[39] Fellay, C.; Yan, N.; Dyson, P. J.; Laurenczy, G. *Chem. - Eur. J.* 2009, 15, 3752.

[40] Himeda, Y. *Green Chem.*2009, 11, 2018.

[41] Fukuzumi, S.; Kobayashi, T.; Suenobu, T. *J. Am. Chem. Soc.* 2010, 132, 1496.

[42] Ritter, S. K. *Chem. Eng. News* 2010, *88*, 26.

[43] Cady, C. W.; Crabtree, R. H.; Brudvig, G. W. *Coord. Chem. Rev.* 2008, 252, 444.

[44] Hetterscheid, D. G. H.; vanderVlugt, J. I.; deBruin, B.; Reek, J. N. H. *Angew. Chem. Int. Ed.* 2009, 48, 8178.

[45] Sala, X.; Romero, I.; Rodríguez, M.; Escriche, L.; Llobet, A. *Angew. Chem. Int. Ed.* 2009, 48, 2842.

[46] Gersten, S. W.; Samuels, G. J.; Meyer, T. J. *J. Am. Chem. Soc.* 1982, 104, 4029.

[47] Gilbert, J. A.; Eggleston, D. S.; Murphy, W. R.; Geselowitz, D. A.; Gersten, S. W.; Hodgson, D. J.; Meyer, T. J. *J. Am. Chem. Soc.* 1985, 107, 3855.

[48] McDaniel, N. D.; Coughlin, F. J.; Tinker, L. L.; Bernhard, S. *J. Am. Chem. Soc.* 2008, 130, 210.

[49] Blakemore, J. D.; Schley, N. D.; Balcells, D.; Hull, J. F.; Olack, G. W.; Incarvito, C. D.; Eisenstein, O.; Brudvig, G. W.; Crabtree, R. H. *J. Am. Chem. Soc.* 2010, 132, 16017.

[50] Savini, A.; Bellachioma, G.; Ciancaleoni, G.; Zuccaccia, C.; Zuccaccia, D.; Macchioni, A. *Chem. Commun.* 2010, 46, 9218.

[51] Lalrempuia, R.; McDaniel, N. D.; Muller-Bunz, H.; Bernhard, S.; Albrecht, M. *Angew. Chem. Int. Ed.* 2010, 49, 9765.

[52] Hetterscheid, D. G. H.; Reek, J. N. H. *Chem. Commun.* 2011, 47, 2712.

[53] Dzik, W. I.; Calvo, S. E.; Reek, J. N. H.; Lutz, M.; Ciriano, M. A.; Tejel, C.; Hetterscheid, D. G. H.; de Bruin, B. *Organometallics* 2011, 30, 372.

[54] Hull, J. F.; Balcells, D.; Blakemore, J. D.; Incarvito, C. D.; Eisenstein, O.; Brudvig, G. W.; Crabtree, R. H. *J. Am. Chem. Soc.* 2009, 131, 8730.

[55] Brewster, T. P.; Blakemore, J. D.; Schley, N. D.; Incarvito, C. D.; Hazari, N.; Brudvig, G. W.; Crabtree, R. H. *Organometallics* 2011, 30, 965.

[56] Blakemore, J. D.; Schley, N. D.; Olack, G. W.; Incarvito, C. D.; Brudvig, G. W.; Crabtree, R. H. *Chem. Sci.* 2011, 2, 94.

[57] Wu, X. F.; Xiao, J. L. *Chem. Commun.* 2007, 2449.

[58] Wu, X.; Wang, C.; Xiao, J. *Platinum Met. Rev.* 2010, 54, 3.

[59] Benyei, A.; Joó, F. *J. Mol. Catal.* 1990, 58, 151.

[60] Ruppert, R.; Herrmann, S.; Steckhan, E. *J. Chem. Soc., Chem. Commun.* 1988, 1150.

[61] Westerhausen, D.; Herrmann, S.; Hummel, W.; Steckhan, E. *Angew. Chem. Int. Ed.* 1992, 31, 1529.

[62] Ogo, S.; Makihara, N.; Watanabe, Y. *Organometallics* 1999, *18*, 5470.

[63] Makihara, N.; Ogo, S.; Watanabe, Y. *Organometallics* 2001, *20*, 497.

[64] Himeda, Y.; Onozawa-Komatsuzaki, N.; Miyazawa, S.; Sugihara, H.; Hirose, T.; Kasuga, K. *Chem. - Eur. J.* 2008, 14, 11076.

[65] Hansch, C.; Leo, A.; Taft, R. W. *Chem. Rev.*1991, 91, 165.

[66] Himeda, Y.; Miyazawa, S.; Onozawa-Komatsuzaki, N.; Hirose, T.; Kasuga, K. *Dalton Trans.* 2009, 6286.

[67] Hashiguchi, S.; Fujii, A.; Takehara, J.; Ikariya, T.; Noyori, R. *J. Am. Chem. Soc.* 1995, 117, 7562.

[68] Fujii, A.; Hashiguchi, S.; Uematsu, N.; Ikariya, T.; Noyori, R. *J. Am. Chem. Soc.* 1996, 118, 2521.

[69] Wu, X. F.; Liu, J. K.; Li, X. H.; Zanotti-Gerosa, A.; Hancock, F.; Vinci, D.; Ruan, J. W.; Xiao, J. L. *Angew. Chem. Int. Ed.* 2006, 45, 6718.

[70] Wu, X. F.; Vinci, D.; Ikariya, T.; Xiao, J. L. *Chem. Commun.* 2005, 4447.

[71] Li, X. H.; Blacker, J.; Houson, I.; Wu, X. F.; Xiao, J. L. *Synlett* 2006, 1155.

[72] Wu, X. F.; Li, X. H.; Zanotti-Gerosa, A.; Pettman, A.; Liu, J. K.; Mills, A. J.; Xiao, J. L. *Chem.- Eur. J.* 2008, 14, 2209.

[73] Wu, X. F.; Li, X. H.; McConville, M.; Saidi, O.; Xiao, J. L. *J. Mol. Catal. A: Chem.* 2006, 247, 153.

[74] Noyori, R.; Yamakawa, M.; Hashiguchi, S. *J. Org. Chem.* 2001, *66*, 7931.

[75] Yamakawa, M.; Ito, H.; Noyori, R. *J. Am. Chem. Soc.* 2000, 122, 1466.

[76] Ma, Y.; Liu, H.; Chen, L.; Cui, X.; Zhu, J.; Deng, J. *Org. Lett.* 2003, *5*, 2103.

[77] Wu, J. S.; Wang, F.; Ma, Y. P.; Cui, X. C.; Cun, L. F.; Zhu, J.; Deng, J. G.; Yu, B. L. *Chem. Commun.* 2006, 1766.

[78] Li, L.; Wu, J.; Wang, F.; Liao, J.; Zhang, H.; Lian, C.; Zhu, J.; Deng, J. *Green Chem.* 2007, 9, 23.

In: Iridium: Occurrence, Characteristics and Applications ISBN: 978-1- 62081-681-3
Editors: Chieko Fukui and Maemi Ono

Chapter 3

IRIDIUM CATALYSTS FOR SUSTAINABLE PROCESSES: GLYCEROL DEHYDROGENATION, ALKYNE POLYMERIZATION AND POST-POLYMERIZATION OF POLYKETONES

Erica Farnetti [1] ***and Corrado Crotti*** [2]
[1]Dipartimento di Scienze Chimiche e Farmaceutiche,
Università di Trieste, Italy
[2]C.N.R. – Istituto Struttura della Materia,
Unità staccata di Trieste, Trieste, Italy

ABSTRACT

The applications of iridium catalysts in green chemistry here described include reactions which span from use of a renewable feedstock such as glycerol, to the development of polymerization and post-polymerization reactions, which are excellent examples of atom-economical processes. A novel glycerol valorization route via selective dehydrogenation takes advantage of iridium-based catalysts which prove to be effective also in glycerol as reaction medium; moreover, alternative routes for selective glycerol transformation can proceed *via* intermediate formation of glycerol acetals or ketals, which is also promoted by organoiridium derivatives with high selectivity. On the other hand, the development of regio and stereoselective alkyne oligo and polymerization provides the corresponding polyacetylenes, or in alternative the cyclotrimerization products. Finally, the iridium-catalyzed transfer hydrogenation of polyketones to the corresponding polyalcohols is described, which represents the first example of catalytic reduction of a polyketone.

INTRODUCTION

It is generally acknowledged that one of the major challenges of modern chemistry is the development of catalytic processes to replace conventional stoichiometric methodologies. As a matter of fact, use of the latter procedures is responsible for the formation of unacceptable amounts of waste. Waste prevention is the first of the Principles of Green Chemistry as formulated by Anastas and Kirchhoff [1]; a quantification of process sustainability under the point of view of waste produced was introduced by Sheldon, who defined the novel parameters of Environmental Factor (E- Factor) and Atom Economy [2]. For such parameters, ideal values of zero and 100%, respectively, can be actually approached by using clean catalytic methodologies. Moreover, catalysis generally implies milder experimental conditions and therefore lower energy input on one hand, and on the other higher process selectivity with consequent lower reagent use and possibility of avoiding product separation procedures. Among the transition metals which appear especially promising for catalytic applications, iridium has emerged in the last years as a most versatile element, as several of its derivatives have provided impressive results in various reactions, such as hydrogenation, transfer hydrogenation, alkane dehydrogenation, hydroformylation, carbonylation and, more recently, water splitting to hydrogen and oxygen [3, 4]. Organoiridium derivatives have been successfully employed also in our lab, especially for hydrogenation and hydrogen transfer reduction of common organic substrates such as ketones and aldehydes, olefins, imines etc. [5]. In the last years our attention focussed on the development of environmentally friendly processes, with iridium derivatives being our preferred choice as catalysts. Our interests span from oligo and polymerization to post-polymerization, and more recently to catalytic reactions for glycerol valorization. Here we review our results in these fields, which prove once more the versatility of iridium in homogeneous catalysis.

RESULTS

Part 1. The Glycerol Challenge: How Can We Make that Blob Valuable?

As side effect of the expanding manufacture of biodiesel fuel we are observing an increase of the availability of glycerol, due to a process which generates about 10% weight of glycerol as byproduct of methanol transesterification of seed oils. At present, the large surplus of glycerol in association with its high functionalization makes it one of the most promising platform chemicals of the near future. However, the very reason that makes this molecule an attractive candidate for chemical reactions, *i.e.* its high degree of functionalization, represents a severe limit to its selective transformation: as a matter of fact, the most challenging problem concerning glycerol valorization is represented by selectivity. Moreover, if on one hand glycerol is a very abundant, safe organic block, on the other its physico-chemical properties such as strong hydrophilicity and high viscosity greatly increase the experimental difficulties.

Under the influence of suitable catalysts, glycerol can undergo a large number of chemical transformations to products of commercial value. Among the most promising processes under current investigation are hydrogenolysis leading to 1,2 or 1,3-propanediol, reforming to CO and H_2, carboxylation to give glycerol carbonate, dehydration to acrolein and oxidation [6]. Glycerol oxidation can lead to a large variety of products, many of which have commercial relevance. In most cases, however, the process selectivity is far from being satisfactory: therefore, the development of catalytic processes which afford a single oxidation product is highly desirable.

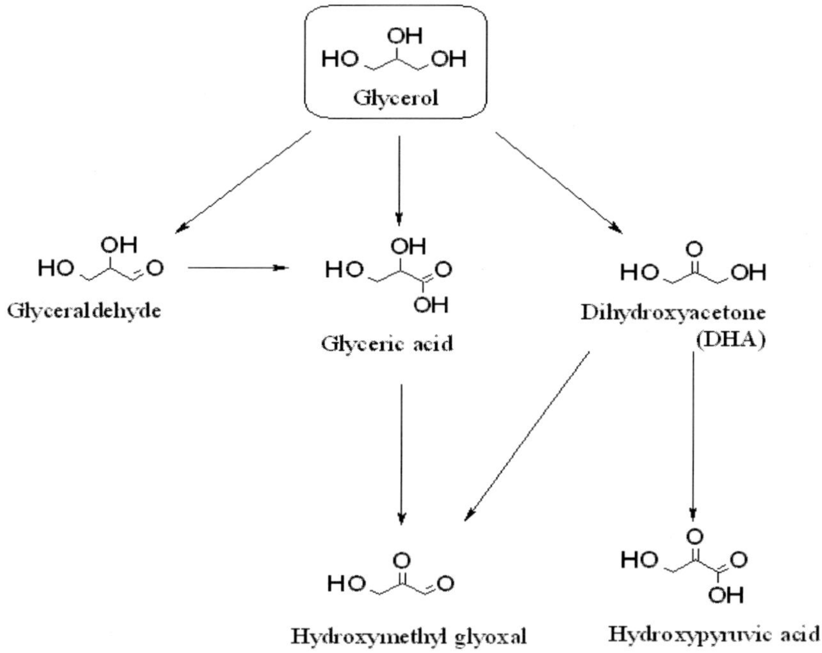

Scheme 1.1.

An efficient catalyst should promote dehydrogenation of either the primary hydroxyl group to produce glyceraldehyde and glyceric acid, or the secondary alcohol function, to give dihydroxyacetone (DHA) and hydroxypyruvic acid (see Scheme 1.1). Presently, DHA is employed in the cosmetic industry as a component in artificial tanning preparations, and when available at lower market price it would find various applications as a versatile building block. Previously reported examples of glycerol oxidation to DHA include microbial fermentation with *Gluconobacter oxydans* [7], which is the currently used method for industrial production, and electrocatalytic oxidation in the presence of TEMPO [8]; heterogeneously catalyzed reactions are generally poorly selective [9], with the exception of some platinum catalysts developed by Kimura and coworkers [10]. Very recently, selective oxidation of glycerol to DHA homogeneously catalyzed by palladium complexes was reported [11], but the catalytic system needs a quite high catalyst/substrate ratio and it suffers for rather fast catalyst deactivation. An efficient, selective catalytic procedure for glycerol oxidation to DHA is presently lacking.

2.1. Dehydrogenation of Glycerol via Hydrogen Transfer Reaction

A possible environmentally friendly route to glycerol oxidation is represented by dehydrogenation *via* hydrogen transfer reaction: glycerol can be envisaged as a disubstituted 2-propanol, therefore it is expected to undergo transfer dehydrogenation in the presence of a suitable catalyst and an appropriate hydrogen acceptor. According to the relative oxidation potentials the hydrogen transfer is expected to selectively occur from the secondary hydroxyl group, yielding DHA as dehydrogenation product. Iridium complexes with nitrogen chelating ligands are well known transfer hydrogenation catalysts for the reduction of various unsaturated substrates [12]. Therefore, we employed these complexes as catalysts to promote the transfer dehydrogenation of glycerol to DHA, using acetophenone as hydrogen acceptor (see Scheme 1.2) [13].

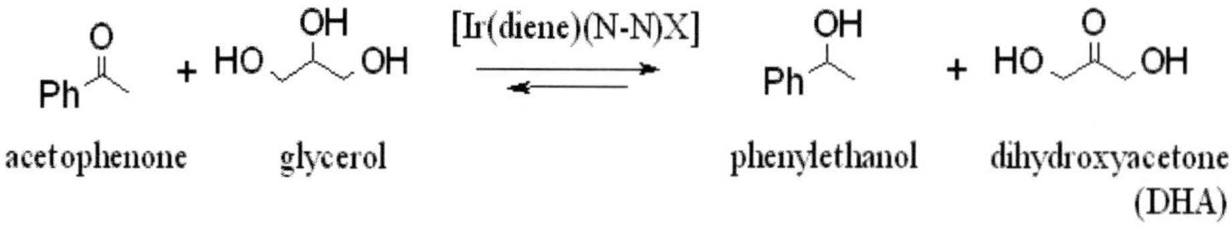

Scheme 1.2.

A first series of reactions was performed using [Ir(diene)(N-N)X] (diene = 1,5-hexadiene (hd) or 1,5-cyclooctadiene (cod); N-N = 3,4,7,8-tetramethyl-1,10-phenanthroline (Me_4phen) or 4,4'-dimethyl-2,2'-bipyridine (Me_2bipy); X = Cl or I) and a basic cocatalyst generally necessary in transfer hydrogenation: the analysis of the reaction mixture revealed that glycerol was behaving as hydrogen donor, partially reducing acetophenone to phenylethanol. The use of glycerol as solvent evidenced some difficulties concerning solubility of catalysts; however, such choice looked coherent with the green chemistry principles.

Among the catalysts tested, [Ir(hd)(Me_4phen)Cl] resulted to be the most efficient one in promoting the hydrogen transfer reaction from glycerol to acetophenone (see a selection of results in Table 1.1). The presence of a basic cocatalyst was necessary in all cases and a significant correlation between cocatalyst basicity and conversion was observed, *i.e.* the lower the K_b value, the lower the conversion of acetophenone to phenylethanol. On the other hand, the selectivity in DHA formation did not follow an analogous correlation upon the strength of the base: apparently the higher the conversion, the lower the DHA selectivity.

Table 1.1. Hydrogen transfer reaction between glycerol and acetophenone catalyzed by [Ir(diene)(N-N)Cl][a]

Entry	Catalyst	Conv. (%)[b]
1	[Ir(hd)(Me_4phen)Cl]	26
2	[Ir(cod)(Me_4phen)Cl]	4
3	[Ir(hd)(Me_2bipy)Cl]	25
4	[Ir(cod)(Me_2bipy)Cl]	6

[a] Experimental conditions: solvent = glycerol; [Ir] = 3.0×10^{-3} mol L^{-1}; [acetophenone]/[Ir] = 100; T = 100°C; reaction time = 1 h; base = NaOH; [base]/[Ir] = 2. [b] calculated as % acetophenone reduced to phenylethanol.

The same conclusions were evidenced for the dependance of conversion and DHA selectivity upon the basic cocatalyst concentration, *i.e.* the [base]/[Ir] ratio. Best results were obtained by the use of K_2CO_3 as cocatalyst, with a [base]/[Ir] ratio of 2. The reaction was very selective, giving DHA as the only glycerol dehydrogenation product, however its relative amount was always lower than that of phenylethanol; moreover, at longer reaction times a notable decrease in DHA yield was monitored (Table 1.2). Further studies confirmed that, indipendently from the presence of catalyst in the reaction medium, DHA was unstable in the basic conditions, which caused an effective loss in the final yield: for example, in catalytic reactions catalyzed by [Ir(hd)(Me_4phen)Cl] and two equivalents of K_2CO_3 at 100°C, in the first 15 minutes the conversion measured as phenylethanol yield was 26%, but the selectivity in DHA had already decreased to 53% mainly due to base-promoted condensation reactions (see Table 1.2, entries 1 and 2, respectively). Analogous effect was observed by increasing the temperature, as shown by comparison of entries 3, 6 and 7.

Table 1.2. Hydrogen transfer reaction between glycerol and acetophenone catalyzed by [Ir(hd)(Me_4phen)Cl][a]

Entry	Reaction Time (min)	T (°C)	Conv. (%)[b]	DHA Yield (Sel.) (%)[c]
1	5	100	15	11 (77)
2	15	100	26	13 (53)
3	30	100	28	13 (47)
4	60	100	33	13 (39)
5	180	100	34	12 (34)
6	30	110	37	14 (37)
7	30	120	47	12 (24)

[a] Experimental conditions: solvent = glycerol; [Ir] = 3.0×10^{-3} mol L^{-1}; [acetophenone]/[Ir] = 100; T = 100 °C; base = K_2CO_3; [base]/[Ir] = 2. [b] calculated as % acetophenone reduced to phenylethanol. [c] DHA yield %: calculated on the basis of the initial acetophenone; Sel. %: calculated on the basis of the Conversion.

2.2. Dehydrogenation of Glycerol via Baseless Hydrogen Transfer Reaction

In order to overcome the difficulty represented by side reactions with the basic cocatalyst, other hydrogen transfer catalysts which are active in baseless conditions had to be considered, which restricted the choice to a very limited number of them [14, 15].

We tested the most promising iridium derivatives according to the literature data and we found that [HIr(cod)(PNP)] (PNP = Pr^n-$N(CH_2CH_2PPh_2)_2$), which had been studied in our laboratories some years ago for the hydrogen transfer reduction of ketones in baseless conditions [16], was the only one among those tested to be active in the transfer dehydrogenation of glycerol [17]; a selected series of results using a variety of hydrogen acceptors is reported in Table 1.3.

The results show that in almost all cases we obtained only small amounts of hydrogenated acceptors and DHA, but the use of benzaldehyde as acceptor gave a much higher conversion to the corresponding benzyilc alcohol and good yields of DHA, together with significative amounts of the acetals formed by condensation of glycerol with benzaldehyde. Notably, aldehydes are not generally viewed as effective hydrogen acceptors [18], on one hand because they easily undergo decarbonylation which produces catalytically inactive species; on the other due to their low stability in the basic conditions required in most cases by hydrogen transfer catalysis. The DHA yields obtained in these reactions, although not excellent, are among the highest so far reported.

2.3. *Acceptorless Dehydrogenation of Glycerol*

Notably, in most of the reactions reported in Table 1.3 the yield of DHA was higher than that of the hydrogenated acceptor: further experiments revealed that two parallel reaction paths are operative, *i.e.* a hydrogen transfer between glycerol and the acceptor and a dehydrogenation of glycerol without the participation of the hydrogen acceptor (see Scheme 1.3). Such acceptorless dehydrogenation of alcohols is an uncommon reaction, although a few examples were reported recently [19], and it doubtless is an extremely attractive process both in terms of atom economy and E-factor [2, 20].

Table 1.3. Hydrogen transfer reaction between glycerol and various acceptors catalyzed by [HIr(cod)(PNP)][a]

Entry	Acceptor	Yield (%)[b]	
		Hydrogenated acceptor	DHA
1	Acetophenone	3	5
2	Cyclohexanone	4	5
3	Styrene	6	8
4	*trans*-Ethylcrotonate	<1	2
5	Benzaldehyde	37	23

[a] Experimental conditions: solvent = glycerol; [Ir] = 3.0×10^{-3} mol L^{-1}; [acceptor]/[Ir] = 100; T = 100°C; reaction time = 1 h. [b] Based on initial acceptor amount.

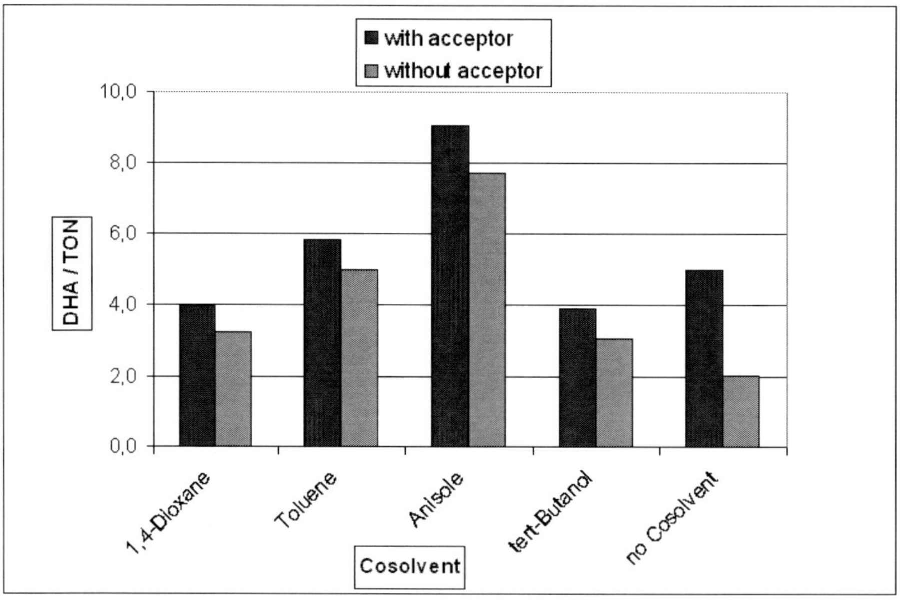

Figure 1.1. Comparison of DHA yields with and without acetophenone as hydrogen acceptor, in the presence of a cosolvent.

In the presence of a suitable cosolvent to increase the catalyst solubility, the acceptorless dehydrogenation proceeds with improved although limited yields, thus demonstrating the feasibility of the greenest possible glycerol selective dehydrogenation.

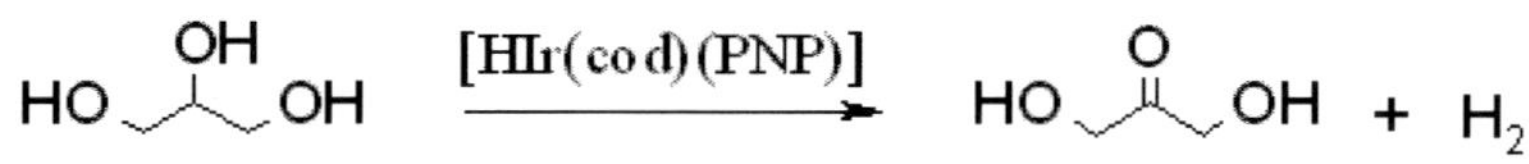

Scheme 1.3.

A comparison of DHA yields (TONs) with and without a hydrogen acceptor, in the presence of various cosolvents is shown in Figure 1.1; the data clearly suggest the simultaneous action of two different pathways – the acceptorless dehydrogenation and the hydrogen transfer reaction – leading to the overall yield of DHA.

2.4. Acetalization of Glycerol

An alternative approach to selective glycerol dehydrogenation may proceed *via* previous protection of two hydroxyl groups by acetalization. As preliminary step we studied glycerol acetalization with various ketones and aldehydes, employing both conventional acid and transition metal-catalyzed procedures. Our results indicated that although acid promoted glycerol acetalization proceeded with good yields in most cases, the reaction selectivities were often moderate or low; on the other hand, late transition metal complexes have been reported to catalyze acetalization of several alcohols and diols with good results [21].

We found that the iridium complexes $[Cp^*IrCl_2]_2$ (Cp* = pentamethylcyclopentadienyl) and $[Cp^*Ir(Bu_2\text{-}NHC)Cl_2]$, ($Bu_2$-NHC = 1,3-di-*n*-butylimidazolylidene) catalyze glycerol acetalization with a series of aldehydes and ketones (see Scheme 1.4) [22].

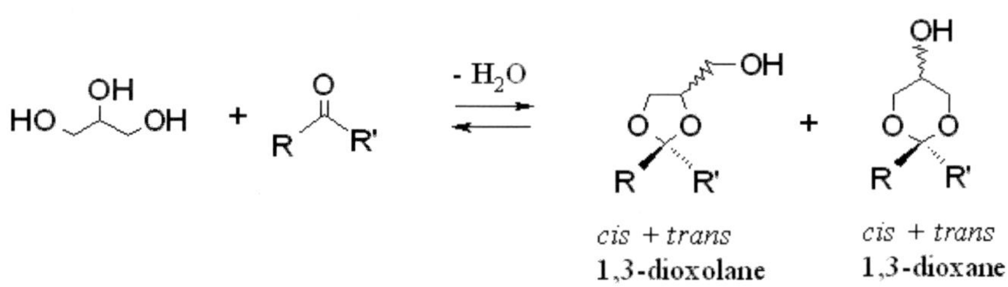

Scheme 1.4.

The reaction proceeds with good to excellent selectivities towards the formation of the 5-membered ring, 1,3-dioxolane (see Table 1.4). The reaction showed a clear dependence on the bulkiness of the carbonyl group substituents, as indicated by the high conversions obtained with acetone and cyclohexanone on one hand, and the lower yields produced with dimethylpentanone on the other; no reaction was observed with benzophenone. The catalytic system was not sensitive to air and rather robust as well, allowing several catalytic cycles with fresh loads of carbonylic compounds. Since the reaction is an equilibrium, the increasing concentration of water in the rection medium affects the reaction yield; thus, we found that the presence of molecular sieves may give a moderate beneficial effect, by shifting the acetalization equilibrium towards the condensation product.

The same catalysts promoted also transacetalization reactions, *i.e.* replacement of acetone with 2, 2-dimethoxypropane leads to the formation of methanol and the same cyclic acetals shown in Scheme 1.4. Also for this reaction the results were very satisfactory (see Table 1.4, entry 7); moreover, we found that catalytic reactions with higher loading of 2,2-dimethoxypropane led to high conversion with resulting TON values exceeding 1400.

Table 1.4. Acetalization of glycerol with various carbonylic compounds catalyzed by $[Cp^*IrCl_2]_2$[a]

Entry	Carbonylic Compound	T (°C)	Conv. (%)[b]	Select. (%)[c]
1	Acetone	50	75	98
2	Acetophenone	120	48	95
3	Cyclohexanone	60	76	97
4	2,4-Me_2-3-pentanone	100	14	100
5	Benzaldehyde	40	55	79
6	Octanal	60	87	82
7[d]	2,2-Dimethoxypropane	40	89	94

[a] Experimental conditions: solvent = glycerol; [Ir] = 3.0×10^{-3} mol L^{-1}; [carbonylic compound]/[Ir] = 500; reaction time = 1 h. [b] Based on initial carbonylic compound amount. [c] Selectivity in terms of 1,3-dioxolane yield. [d] Transacetalization reaction.

This efficient glycerol acetalization procedure has some applicative importance because not only glycerol cyclic acetals and ketals are themselves valuable chemicals, but they could be also employed as new substrates: on one hand dehydrogenation of the free hydroxyl group of the 1,3-dioxolane would produce the corresponding aldehyde, on the other the 1,3-dioxane would give the corresponding ketone (see Scheme 1.5). Final deprotection leads to glyceraldehyde and DHA, respectively.

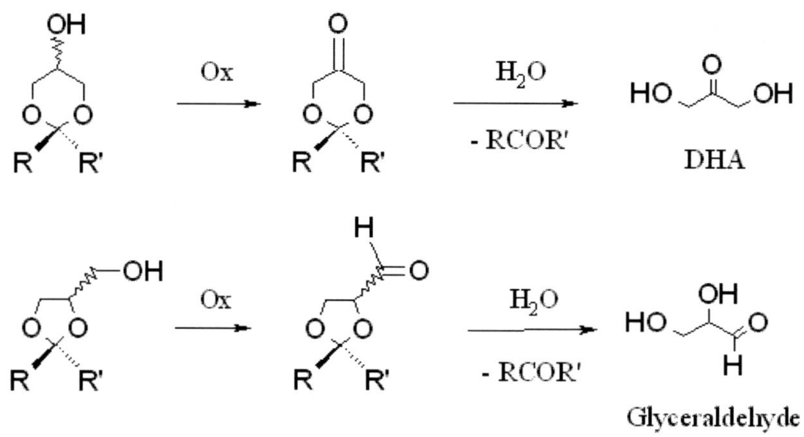

Scheme 1.5.

Part 2. Iridium-Catalyzed Polymerization and Post-polymerization

Few processes are known which are characterized by ideal values of atom utilization and E-factor, the most typical examples being polymerization processes. As a matter of fact efficient, selective conversions of one or more monomers to a polymeric (or oligomeric) products are accompanied by little or no waste formation. Homogeneously catalyzed polymerization of unsaturated monomers offers these advantages in terms of process sustainability on one hand, and on the other such process can provide new high value polymeric materials with a variety of interesting properties. Moreover, in recent years an alternative approach has been developed, which regards the modulation of physical and chemical properties of a known polymeric material by chemical modification via post-polymerization. This approach has among the possible targets that of transforming a non-degradable polymer into a biodegradable material, once more to meet environmental issues. Also this target is best approached by use of efficient, selective catalysts.

Among the polymers which have recently attracted attention are polyacetylenes, which are considered to be promising materials due to their physico-chemical properties (*e.g.* photoluminescence, photoconductivity, oxygen permeability) [23-26]. Such polymers can be obtained by homogeneously catalyzed polymerization of terminal alkynes, which generally occurs with excellent head to tail regiochemistry (Scheme 2.1). With regard to stereochemistry, four possible isomers can be obtained which differ for configuration of the C=C bond and/or conformation of the C-C bond (see Figure 2.1). According to the choice of the catalytic system, the polyene can be obtained with prevalent *trans* or *cis* stereochemistry. Moreover, during the catalytic reactions formation of variable amounts of oligomeric products (dimers and cyclotrimers) can be observed, all of which are useful synthons for organic synthesis.

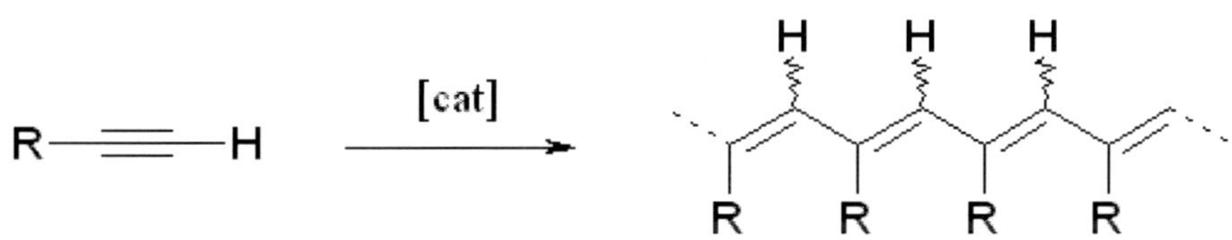

Scheme 2.1.

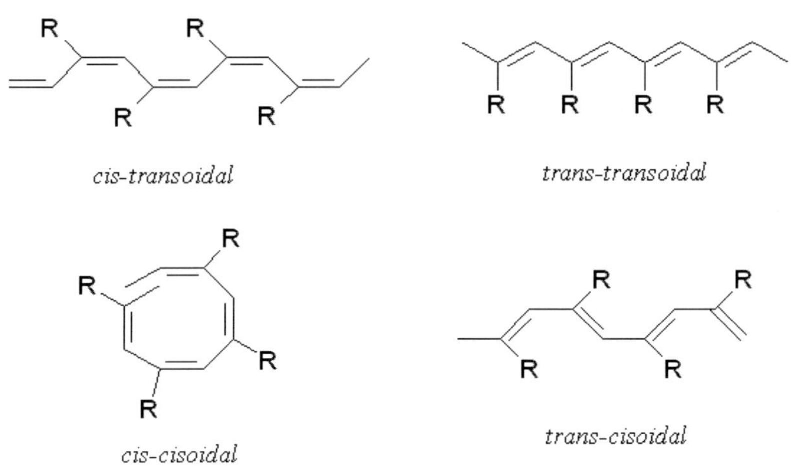

Figure 2.1. Stereoisomers of polyacetylenes.

The catalytic systems which are known to promote alkynes polymerization can be divided into two main groups, *i.e.* rhodium derivatives on one hand, and molybdenum and tungsten-based systems on the other. The first group includes several organorhodium species, some of which behave in a living fashion: all such catalysts promote the polymerization of terminal alkynes to the corresponding *cis*-polyenes, with high stereoselectivities [27-33].

The reaction occurs via an insertion mechanism, as elegantly demonstrated by Noyori and coworkers [29]. The catalysts of the second group, which are mainly molybdenum derivatives, generally produce the *trans*-polyenes as major isomer, although the selectivity is very variable: in this case a metathesis mechanism *via* a metal-carbene active species appears to be operative [34-37].

In view of the excellent properties of rhodium-based catalysts, also iridium derivatives appeared to be promising candidates: therefore our group started an investigation on organoiridium species as catalysts for alkynes polymerization.

2.1. Alkynes Polymerization Catalyzed by [Ir(diene)X]$_2$ (X=Cl, OMe)

The dimeric derivatives [Rh(diene)Cl]$_2$ (diene= cod or 2,5-norbornadiene) were among the first catalytic systems which were studied for phenylacetylene polymerization [38]; more recently, in our laboratories we tested the methoxy species [Rh(cod)(OMe)]$_2$ which proved to be more efficient than the corresponding chloro dimers [39].

The iridium analogues [Ir(cod)X]$_2$ (X=Cl, OMe) (Figure 2.2) proved to be also active for the polymerization of phenylacetylene, in experimental conditions similar to those employed for rhodium-catalyzed reactions (*e.g.* solvent tetrahydrofuran, 60°C), although the final conversion never exceeded 50% due to catalyst deactivation (see Table 2.1, entries 2-5) [40]. The presence of coordinated diene is an essential requirement for polymerization catalysis, as demonstrated by comparison of entries 1 and 2 in Table 2.1: when cod is substituted with two monoenes (coe = cyclooctene) no polymer is formed, at variance mixtures of cyclotrimerization products are detected. Interestingly, in contrast to the results reported for rhodium catalysts, which always give the *cis*-polyene as main product, the iridium dimers promote the formation of polyacetylenes with a configuration which is dependent on the reaction temperature: at r.t. the *cis*-polyene is obtained, whereas al 60°C the polyphenylacetylene has mainly *trans* configuration (in Table 2.1 compare entries 2,3 and 4,5).

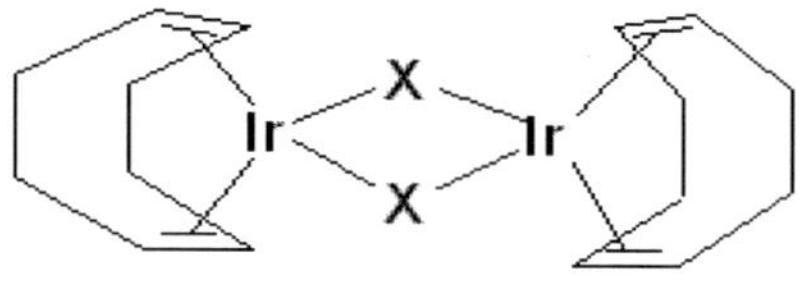

X = Cl, OMe

Figure 2.2.

The evolution of the iridium dimer during the catalytic reactions was studied by NMR spectroscopy, using experimental conditions similar to those of the catalytic reactions apart from catalyst and substrate concentrations. The results of these studies gave several hints on the fate of the catalyst precursor, in particular the presence of free cod in amounts increasing with time indicates that catalyst deactivation is probably related to loss of the diene from the coordination sphere of iridium.

According to the hypothesis that the catalytically active species was probably formed *via* solvent induced splitting of the dimeric compound into monomeric species, more coordinating solvents were expected to provide an improvement on the catalytic reaction. In fact, use of triethylamine as reaction medium resulted in higher conversions, although catalyst deactivation was still occurring and therefore limited the final yields of polyphenylacetylene (see Table 2.1, entries 6,7 and 9,10) [40].

Interestingly, the increased apparent catalytic activity, probably related to higher concentration of the initiator, allowed to perform the reactions at temperatures as low as 0°C. The effect of reaction temperature on polyene stereochemistry can be appreciated by comparing entries 6, 7, 8 and 9, 10, 11 of Table 2.1: the *cis*:*trans* content reverts from about 25:75 at 60°C to 75:25 at 0°C.

Table 2.1. Polymerization of phenylacetylene catalyzed by $[Ir(cod)X]_2$[a]

Entry	Catalyst	Solvent	T (°C)	Conv. (%)	*trans*-PPA[b] (%)	*cis*-PPA[b] (%)	triphenyl-benzenes[c] (%)
1	$[Ir(coe)_2Cl]_2$	THF	60	10	0	0	100
2	$[Ir(cod)Cl]_2$	THF	60	45	71	27	2
3	$[Ir(cod)Cl]_2$	THF	25	20	5	95	-
4	$[Ir(cod)(OMe)]_2$	THF	60	51	75	21	4
5	$[Ir(cod)(OMe)]_2$	THF	25	36	2	93	5
6	$[Ir(cod)Cl]_2$	NEt_3	60	57	75	25	-
7	$[Ir(cod)Cl]_2$	NEt_3	25	44	32	68	-
8	$[Ir(cod)Cl]_2$	NEt_3	0	46	26	74	-
9	$[Ir(cod)(OMe)]_2$	NEt_3	60	61	76	24	-
10	$[Ir(cod)(OMe)]_2$	NEt_3	25	45	46	64	-
11	$[Ir(cod)(OMe)]_2$	NEt_3	0	48	25	75	-

[a] Experimental conditions: [Ir] = 8.0×10^{-3} mol L^{-1}; [sub]/[Ir] = 50; reaction time: 5 h. [b] PPA = polyphenylacetylene. [c] 1,3,5 and 1,2,4-triphenylbenzene.

The reaction temperature had a remarkable effect also on the polyene molecular weights which generally increase when decreasing the reaction temperature, with M_n values ranging from about 3500 at 60°C to 7000 at 0°C, with polydispersion index values (M_w/M_n) between 1.4 and 1.7.

Spectroscopic studies performed on the catalytic systems using $[Ir(cod)Cl]_2$ and NEt_3 as solvent evidenced formation of the monomeric species $[Ir(cod)(NEt_3)Cl]$, thus supporting the hypothesis that splitting of the bridging ligand of the dimer by a solvent molecule is a key step for the formation of the polymerization initiator.

2.2. Alkynes Polymerization Catalyzed by [Hir(diene)(PR$_3$)$_2$]

Derivatives of the type [HIr(diene)(PR$_3$)$_2$] can be synthezised by reacting [Ir(diene)(OMe)]$_2$ with 2 equivalents of the chosen phosphine. The reaction involves preliminary formation of the phosphine-methoxo compound, which undergoes hydrogen β-elimination to the corresponding hydrido species and free formaldehyde (see Scheme 2.2). Notably, whereas with phosphines having intermediate cone angles (*e.g.* PPh$_3$, PCyPh$_2$, P(*p*-MeOC$_6$H$_4$)$_3$) the reaction readily proceeds to formation of the hydrido compound, with bulkier phoshines (*e.g.* PCy$_3$ and P(*o*-MeOC$_6$H$_4$)$_3$ the reaction stops at the methoxo derivatives without undergoing β-elimination, probably because formation of the hydride-formaldehyde intermediate is prevented in the too crowded coordination sphere.

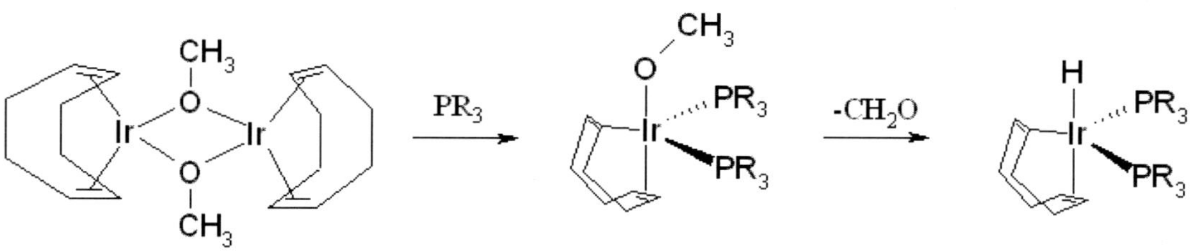

Scheme 2.2.

Both the methoxy and the hydride derivatives were tested as catalysts for phenylacetylene polymerization: the former complexes only gave poor results, whereas the hydride species were active in both polymerization and dimerization reactions [41]. In a typical catalytic reaction performed in the presence of [HIr(cod)(PPh$_3$)$_2$] and phenylacetylene as monomer, total yields up to 100% of monomer consumption were observed, with important variations in the products composition as a function of solvent and reaction temperature.

Selected results reported in Table 2.2 evidence that in almost all cases the major product was polyphenylacetylene, which was accompanied by variable amounts of the dimerization products (E)-1,4-diphenylbut-1-yn-3-ene and (Z)-1,4-diphenylbut-1-yn-3-ene (see Figure 2.3) and, in some cases, by traces of the two cyclotrimerization products 1,3,5 and 1,2,4-triphenylbenzene. A closer analysis of the reported data reveals that an increase of the reaction temperature favours the dimerization path at the expences of the polymerization process (see Table 2.2, entries 5-7).

Interestingly, in all cases a highly stereoregular *trans*-polyene was obtained; molecular weight determination by GPC gave M_n values between 3000 and 5500, with rather low polydispersion index (M_w/M_n = 1.4-1.6).

A critical analysis of the data reported in Table 2.2 suggests that two different catalytic paths can be activated, the polymerization process being the prevalent one at lower temperatures, whereas the dimerization reaction predominates at higher temperatures.

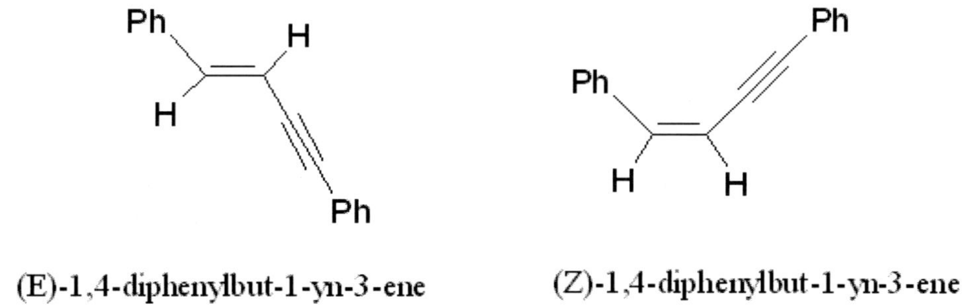

Figure 2.3. Phenylacetylene dimerization products.

Table 2.2. Polymerization of phenylacetylene catalyzed by [HIr(cod)(PR$_3$)$_2$][a]

Entry	PR$_3$	Solv.	T (°C)	Conv. (%)	*trans*-PPA[b] (%)	Dimers[b] (%)
1	PCyPh$_2$	THF	60	68	90	10
2	P(p-MeOC$_6$H$_4$)$_3$	THF	60	44	90	8
3	P(o-MeOC$_6$H$_4$)Ph$_2$	THF	60	39	75	15
4	PPh$_3$	THF	60	47	91	9
5	PPh$_3$	toluene	60	44	75	25
6	PPh$_3$	toluene	80	79	55	45
7	PPh$_3$	toluene	100	100	39	61

[a] Experimental conditions: [Ir] = 3.4×10^{-3} mol L^{-1}; [sub]/[Ir] = 100; reaction time: 5 h. [b] Selectivity (%); PPA = polyphenylacetylene; dimers = (E)-1,4-diphenylbut-1-yn-3-ene and (Z)-1,4-diphenylbut-1-yn-3-ene.

It was reasonable to suppose that two different catalytic species were responsible for the two reaction routes. In order to favour alkyne polymerization lower temperatures appear to be preferable, however at temperatures lower than 60°C a drop of reaction yields was observed. An additional problem was represented by catalyst deactivation, which apparently shortened the lifetime of the polymerization initiator. Such deactivation was initially ascribed to loss of the diene from iridium, in analogy to the results obtained with iridium-cod dimers. However, it was possible to rule out this deactivation route as on one hand no free cod was detected in the final reaction mixtures, and on the other no effect was observed when catalytic reactions were performed in the presence of added diene. Further attempts to provide evidence for the identification of the catalytically active species as well as the nature of the deactivation process led to the isolation of a reaction intermediate, namely [HIr(C≡CPh)$_2$(PPh$_3$)$_3$]. This species proved to be catalytically inactive, and so were the corresponding compounds with PCyPh$_2$, P(*p*-MeOC$_6$H$_4$)$_3$. On the other hand, useful information was acquired by monitoring the development or the catalyst precursor in the presence of the alkyne, by means of infrared and ^{1}H, ^{13}C and ^{31}P NMR spectroscopy. The results allowed us to ascertain that evolution of [HIr(cod)(PPh$_3$)$_2$] followed the two paths reported in Scheme 2.3.

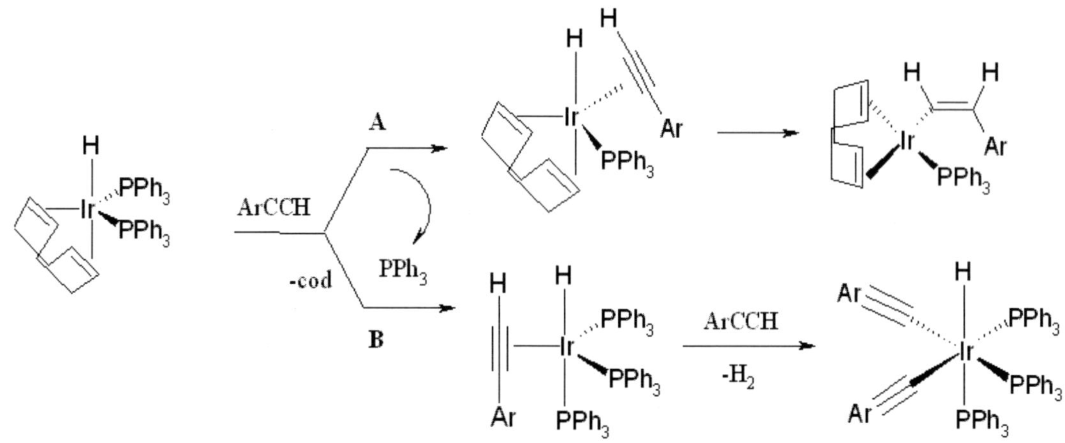

Scheme 2.3.

In path A, initial formation of [HIr(cod)(HC≡CPh)(PPh_3)] via substitution of one phosphine by one monomer molecule was followed by alkyne insertion into the Ir-H bond to give the vinyl compound [Ir(cod)(CH=CHPh)(PPh_3)]. On the other hand, the fate of the catalyst precursor described in path B includes displacement of the diene by the phosphine released in path A, followed by coordination of the alkyne to give the intermediate [HIr(HC≡CPh)$(PPh_3)_3$]; alkyne C-H oxidative addition, H_2 reductive elimination, and finally oxidative addition of a second molecule of phenylacetylene leads to the isolated bis-alkynyl compound [HIr$(C≡CPh)_2(PPh_3)_3$]. Of the two reaction routes, one (path A) leads to formation of the iridium species [Ir(cod)(CH=CHPh)(PPh_3)] which is likely to be the polymerization initiator, whereas the other (path B) represents the catalyst deactivation process.

2.3. Alkynes Cyclotrimerization Catalyzed by Iridium Complexes with Diphosphines

Bidentate phosphines have been extensively employed in homogeneous catalysis, as they offer several advantages in comparison to monodentate analogues, such as increased stability and control of stoichiometry and geometry, which are important properties for their catalytic applications, especially with regard to selectivity. The initial target was to synthesize the diphosphine analogues of [HIr(diene)$(PR_3)_2$] by reacting [Ir(diene)(OMe)]$_2$ with one equivalent of the bidentate phosphine. By this procedure we performed the preparation of [HIr(cod)(dppe)] (dppe = $Ph_2P(CH_2)_2PPh_2$)), obtaining in nearly quantitative yield an off-white solid. However, spectroscopic data revealed that the isolated complex was not the expected compound but its isomer, in which cod was coordinated in an unusual η^1,η^3 fashion (see Scheme 2.4) [42].

In fact, using the same procedure Oro and coworkers had previously obtained the corresponding species with the ligand dppm (= $Ph_2PCH_2PPh_2$), *i.e.* [HIr(η^1,η^3-cod)(dppm)]. According to the studies reported by Oro on the dppm derivative [43], the rearrangement occurs *via* a succession of olefin insertions into the metal-hydride bond and β-eliminations from the iridium-alkyl. Interestingly, when we performed the synthesis of the iridium derivatives with the bidentate phosphine dppp (= $Ph_2P(CH_2)_3PPh_2$) we obtained the corresponding [HIr(diene)(P-P)] compoud as mixtures of isomers, *i.e.* [HIr(η^2,η^2-cod)(dppp)] and [HIr(η^1,η^3-cod)(dppp)], the former being the major product.

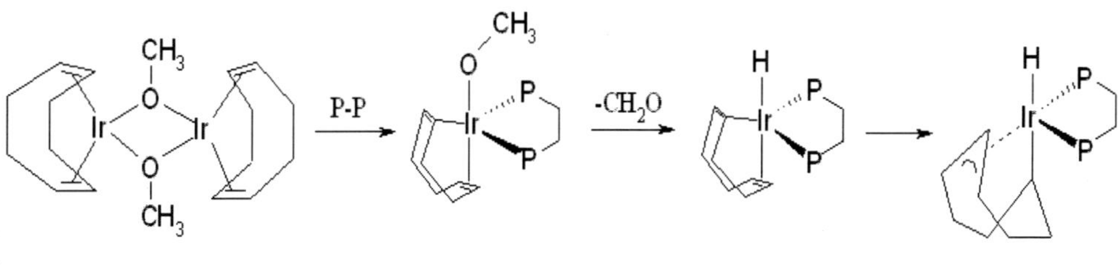

Scheme 2.4.

Finally, reaction of $[Ir(diene)(OMe)]_2$ with dppb (= $Ph_2P(CH_2)_4PPh_2$) gave as only product the derivative with common η^2,η^2-cod coordination, *i.e.* $[HIr(\eta^2,\eta^2\text{-cod})(dppb)]$. Therefore, isomerization of the diene from (η^2,η^2) to (η^1,η^3) coordination mode appeared to be dependent of the iridium-phosphine chelate ring size, being more pronounced with ligands forming the smallest chelate rings.

This hypothesis was strengthened by the results obtained with other diphosphines which also form five-membered ring chelates like dppe, such as o-$C_6H_4(PPh_2)_2$ and $Cy_2P(CH_2)_2PCy_2$ (dcpe): in all cases formation of the single (η^1,η^3) isomer was observed. Moreover, spectroscopic studies indicated that the isomerization η^2,η^2-cod to η^1,η^3-cod was generally favoured by increasing the temperature: for example, the mixture $[HIr(\eta^2,\eta^2\text{-cod})(dppp)]$ and $[HIr(\eta^1,\eta^3\text{-cod})(dppp)]$ changed from 8:1 at 25°C to 1:3 at 55°C, the isomerization process being irreversible.

The iridium diphosphine compounds above discussed were tested as catalyst precursors for phenylacetylene polymerization in THF at 60°C. The rates of the catalytic reactions measured as monomer consumption were markedly dependent on the phosphine employed, decreasing in the series dppm > dppe > dppp > dppb. However, rather surprisingly no polyphenylacetylene was formed, the only products being in all cases the two cyclotrimers 1,3,5 and 1,2,4-triphenylbenzene (Scheme 2.5). Interestingly, the catalytic reactions appear to be highly regioselective [44], as the asymmetric 1,2,4 regioisomer was obtained with selectivities up to 99%, with [HIr(cod)(dppm)] proving both the most active and selective catalyst (see Table 2.3, entries 1-5).

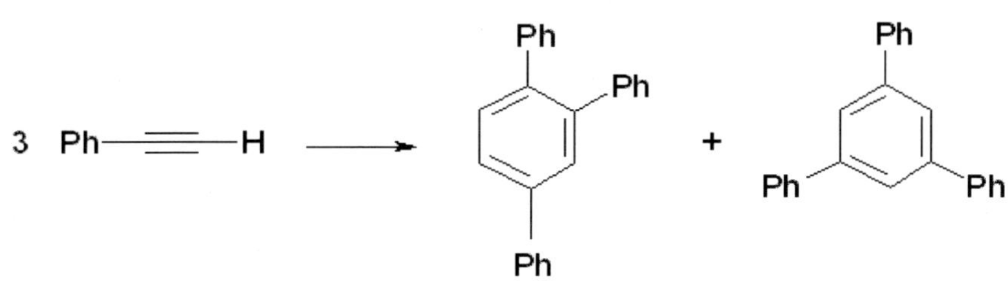

Scheme 2.5.

An exception to this trend was represented by the derivative with dcpe, which only gave moderate conversions to the polyene as the major product, together with small amounts of the cyclotrimers (see Table 2.3, entry 6).

Table 2.3. Cyclotrimerization of phenylacetylene catalyzed by [HIr(cod)(P-P)][a]

Entry	P-P	Conv.(%)	1,2,4-$Ph_3C_6H_3$(%)	1,3,5-$Ph_3C_6H_3$(%)	PPA(%)
1	dppm	100[b]	99	1	-
2	dppe	98	97	1	-
3	dppp	47	45	2	-
4	dppb	31	28	3	-
5	*o*-$C_6H_4(PPh_2)_2$	92	88	4	-
6	dcpe	31	3	3	18

[a] Experimental conditions: [Ir] = 3.4×10^{-3} mol L^{-1}; [sub]/[Ir] = 100; solvent:THF; T=60°C; reaction time: 6 h. PPA = *trans*-polyphenylacetylene. [b] reaction time 15 min. Other reaction products: (E)-1,4-diphenylbut-1-yn-3-ene and (Z)-1,4-diphenylbut-1-yn-3-ene.

The studies of alkyne cyclotrimerization catalyzed by [HIr(cod)(dppm)] were extended to a series of *ortho* and *para*- substituted phenylacetylenes, in order to investigate the effect of electronic and steric monomer variations. The results (see Table 2.4) indicated that the catalytic reactions are favoured by electrodonating substituents, which appear to enhance both reaction rate and regioselectivity, the latter being 100% with the monomer *p*-$MeOC_6H_4C{\equiv}CH$. On the other hand, steric factors cause major effects on the catalytic reactions, as for example with the monomer *o*-$CF_3C_6H_4C{\equiv}CH$ a mixture of polymerization, dimerization and cyclotrimerization products was obtained. These catalytic reactions represent an uncommon example of highly regioselective alkyne cyclotrimerization: although cyclotrimerization of terminal alkynes can be promoted by a variety of transition metals, comparable selectivities were reported in only few cases [45-47].

Table 2.4. Cyclotrimerization of substituted phenylacetylenes catalyzed by [HIr(cod)(dppm)][a]

Entry	Substrate	T (°C)	T (min)	conv. (%)	1,2,4-$Ar_3C_6H_3$ (%)	1,3,5-$Ar_3C_6H_3$ (%)
1	$PhC{\equiv}CH$	60	5	90	89	1
2	*p*-$MeOC_6H_4C{\equiv}CH$	60	5	100	100	-
3	*p*-$CF_3C_6H_4C{\equiv}CH$	60	5	87	86	1
4	*o*-$CF_3C_6H_4C{\equiv}CH$	60	300	85[b]	22	38
5	$PhC{\equiv}CH$	40	90	98	95	3
6	*p*-$MeOC_6H_4C{\equiv}CH$	40	60	100	97	3
7	*p*-$CF_3C_6H_4C{\equiv}CH$	40	120	98	97	1

[a] Experimental conditions: [Ir] = 3.4×10^{-3} mol L^{-1}; [sub]/[Ir] = 100; solvent:THF. [b] Other products: *trans*-PPA, (E)-1,4-diphenylbut-1-yn-3-ene and (Z)-1,4-diphenylbut-1-yn-3-ene.

2.4. Alkynes Polymerization Catalyzed by Iridium Complexes with Nitrogen-Donor Ligands

Organorhodium derivatives with nitrogen chelating ligands such as 2,2'-bipyridine (bipy), 1,10-phenanthroline (phen) and substituted derivatives have long been known to be active catalysts for alkyne polymerization [48]. We were interested in testing the catalytic properties of the iridium analogues in order to provide further evidence on the effect of nature of the ancillary ligands on the selectivity of iridium-promoted reactions: up to that point, iridium oligo and polymerization catalysis had been evidenced to be dramatically influenced by the nature of coordinated ligands, as discussed in the previous paragraphs. Therefore, the complexes [Ir(diene)(N-N)X], (diene = cod, hd; N-N = bipy, phen and substituted derivatives; X = Cl, Br, I) were tested as catalyst precursors for phenylacetylene polymerization [49].

Table 2.5. Polymerization of phenylacetylene catalyzed by [Ir(diene)(N-N)X][a]

Entry	Catalyst	solvent	T (°C)	[NaOH]/[Ir]	time (h)	Conv. (%)
1	[Ir(cod)(phen)Cl]	toluene	110	2	5	23
2	[Ir(cod)(phen)Cl]	toluene	110	-	5	0
3	[Ir(hd)(phen)Cl]	toluene	110	2	5	25
4	[Ir(hd)(phen)Cl]	toluene	110	-	5	31
5	[Ir(cod)(phen)Cl]	mesitylene	160	2	5	58
6	[Ir(cod)(phen)Cl]	mesitylene	160	-	5	46
7	[Ir(hd)(phen)Cl]	mesitylene	160	2	5	46
8	[Ir(hd)(phen)Cl]	mesitylene	160	-	5	73
9	[Ir(hd)(phen)Br]	mesitylene	160	-	5	97
10	[Ir(hd)(phen)I]	mesitylene	160	-	2	98
11	[Ir(hd)(Ph_2phen)Cl]	mesitylene	160	-	10	58
12	[Ir(hd)(phen)Cl]	mesitylene	160	-	10	81
13	[Ir(hd)(Me_2phen)Cl]	mesitylene	160	-	10	86
14	[Ir(hd)(Me_4phen)Cl]	mesitylene	160	-	10	94

[a] Experimental conditions: [Ir] = 6.8×10^{-3} mol L^{-1}; [sub]/[Ir] = 50.

The iridium derivatives proved to be active catalysts as they promoted formation of polyphenylacetylene as the only reaction product: dimers and cyclotrimers were never detected in the final reaction mixtures. The reactions can be carried out in refluxing toluene, however in this solvent final yields were lower than 70% in all cases; use of the higher boiling mesitylene allowed to get close to 100% conversion when performing the catalytic reactions at 160°C. Selected results reported in Table 2.5 summarize the effect of the nature of the various ligands on the catalytic reaction.

Iridium cod-derivatives are generally less active than hexadiene analogues; the presence of a basic cocatalyst produced an increase in reaction yields when using cod-catalysts (see Table 2.5, entries 1-2 and 5-6) and notably, in toluene at 110°C such compounds are totally inactive in baseless conditions (entry 2), but the opposite effect is observed with hd-derivatives (entries 3-4 and 7-8 in Table 2.5), making the latter species more apt to be used with base-sensitive monomers. A second effect was observed when comparing complexes with different nitrogen ligands: as a general trend, the presence of better electron-donating substituents had a beneficial influence on the reaction conversion, as evidenced by a comparison of entries 11-14 in Table 2.5.

Also the choice of the halogen ligands appears to have a significative effect on the catalytic reaction, with the observed trend of final conversion being I > Br > Cl (see Table 2.5, entries 8-10). In all the catalytic reactions performed with the iridium complexes with nitrogen ligands formation of *trans*-polyphenylacetylene was observed, with molecular weight (M_n) values up to 10000.

2.5. Iridium-Catalyzed Post-polymerization of Polyketones

An alternative approach to the synthesis of new polymeric materials can be provided by the chemical modification of known polymers. This approach is attracting increasing attention by the academic and industrial community, as demonstrated by several reports in the recent literature [50]. Especially interesting appear those post-polymerization procedures which strongly modify the physico-chemical properties of the polymer, as in principle it should be possible to control its final properties by careful tuning of the reaction conditions.

Among the polymeric materials which appear to be especially suitable for postpolymerization are polyketones: these polymers have been the object of intensive studies in view of various potential industrial applications. Polyketones can be prepared from carbon monoxide and terminal olefins through palladium-catalyzed strictly alternating copolymerization [51-54] but in spite of their unique properties (high crystallinity, photodegradability), up to now they have not found major applications.

However, the possibility of chemically modifying these polymers by reaction of the carbonyl groups in the polymer backbone offers a possible access to a variety of functionalized polymers. One of the most promising post-polymerization reactions on polyketones is the reduction of carbonylic functions to give the corresponding polyalcohols: such reaction would provide a highly hydrophylic polymeric material; moreover, the polymer hydrophylicity could be controlled by performing a partial or quantitative polyketone reduction.

Very few reports have been published on polyketone reduction methods, all of which make use of stoichiometric reagents such as aluminium or boron hydrides [55, 56]. As these procedures are environmentally unfriendly due to the use of large excesses of reductant, it would be highly desirable to develop an efficient catalytic process; moreover, among the catalytic procedures for ketone reduction, transfer hydrogenation represents the preferable choice in comparison to use of molecular hydrogen.

Therefore, we examined the possible application of iridium-based catalysts for the reduction of polyketones in hydrogen-transfer conditions. The polymeric materials of choice were atactic CO/styrene and CO/4-Me-styrene polyketones, obtained by palladium-catalyzed copolymerization [57], using bis(aryl)acenaphthenequinonediimide ligands in 2,2,2-trifluoroethanol. The atactic polyketones employed (M_n in the range 7–40 $\times 10^3$) are characterized by higher solubility in organic solvents, in comparison to polyketones having a different tacticity.

In fact, substitution of ordinary substrates with macromolecules provides a novel difficulty for the catalytic reaction, namely the polymer solubility, which is generally low in common organic solvents, and even lower in the alcoholic medium generally required for transfer hydrogenation. After several tests it was possible to identify dioxane as the best choice as cosolvent; therefore, an optimized procedure for transfer hydrogenation of polyketones was established, using a 2-propanol/cosolvent ratio of 5:3 which allowed to obtain a compromise between reasonably good solubility of the polymer and relatively high content of hydrogen donor. In this solvent mixture the polymeric material was only partially soluble at r.t., however at the usually employed reaction temperature (83°C) a homogeneous mixture was obtained.

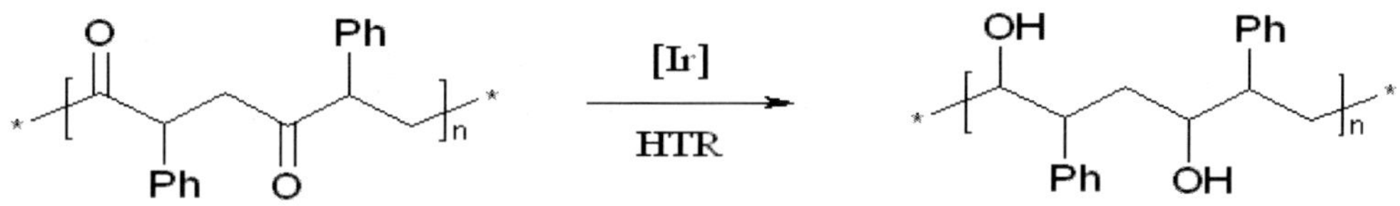

Scheme 2.6.

Thus, hydrogen transfer reduction of CO/styrene polyketone (see Scheme 2.6) was accomplished using the catalyst [Ir(cod(N-N)Cl] (N-N = Me_2bipy, Me_4phen) at 83°C and in the presence of KOH as basic cocatalyst. In a typical catalytic reaction carried out in a misture of 2-propanol and dioxane, conversions higher than 70% were obtained in 8h [58]. Interestingly, substitution of Me_2bipy or Me_4phen with other bipyridine or phenanthroline (*e.g.* bipy, phen, Me_2phen) caused a dramatic decrease of the catalytic activity. Moreover, as previously observed in transfer hydrogenation of other substrates catalyzed by complexes of the type [Ir(cod(N-N)X], in the absence of basic cocatalyst no reaction was observed.

Other catalytic tests were performed with the same catalysts and cocatalyst but using the CO/4-Me-styrene polyketone, giving results either similar or superior to those observed with the unsubstituted polymer. From the selected data reported in Table 2.6 the catalytic reaction appears to be significantly affected by the choice of both nitrogen and halogen ligands.

A perusal of the results reported in the Table reveals that, among the bipyridines and phenanthrolines tested, Me_2bipy and Me_4phen gave rise to the most active catalysts: therefore, the reaction appears to be favoured by strong electron-releasing nitrogen ligands. On the other hand, chloro derivatives appear to be more active than iodo analogues (compare entries 3-7 and 4-6 in Table 2.6), however such trend is expected to be a combination of the effect of donating properties and halogenide dissociation, as the latter partially occurs in the alcohol-dioxane medium employed.

Interestingly, if on one hand at longer reaction times it was possible to obtain polyalcohol yields close to 100% (see Table 2.6, entry 5), on the other by stopping the catalytic process at the appropriate reaction time various intermediate polymeric materials could be produced, with tunable properties as a function of carbonyl/alcoholic group ratio.

Table 2.6. Transfer hydrogenation of CO/4-Me-styrene catalyzed by [Ir(cod)(N-N)X][a]

Entry	N-N	X	t (h)	conv. (%)
1	Me_2phen	Cl	7	0
2	Me_4phen	Cl	7	45
3	bipy	Cl	7	10
4	Me_2bipy	Cl	7	90
5	Me_2bipy	Cl	10	98
6	Me_2bipy	I	7	55
7	bipy	I	7	5

[a] Experimental conditions: [Ir] = 3.0×10^{-3} mol L^{-1}; [repetitive unit]/[Ir] = 100; [KOH]/[Ir] = 2. Hydrogen donor: 2-propanol; cosolvent: dioxane; 2-propanol:cosolvent = 5:3; T = 83°C.

Conclusion

We have described some examples where iridium catalysts can be applied to promote reactions which represent hot issues in the field of environmental and sustainable chemistry. For all cases, control of the chemical and physico-chemical properties of the iridium derivatives allow the fine tuning necessary to meet challenging requirements of selectivity, chemical compatibility, solubility, demanding rection media. In the case of glycerol valorization - a key issue for sustainable biofuel production - selective oxidation can be accomplished in the presence of different iridium catalysts. As a general rule, the representation of glycerol as a disubstituted 2-propanol, and therefore its dehydrogenation *via* hydrogen transfer reaction catalyzed by iridium derivatives such as [Ir(diene)(N-N)X] has been demostrated to be feasible. However the crucial point appears to be the reactivity of the desired product dihydroxyacetone, as its final yields are affected by side reactions with the basic cocatalyst if present, as well as by thermal degradation. To ovecome such problems in the DHA production, the approach here reported spans from use of baseless catalytic systems such as [HIr(cod)(PNP)], to an alternative route which requires previous protection of two hydroxyl groups *via* selective acetalization. In the latter case, after converting glycerol to the corresponding 5- or 6-membered cyclic acetal or ketal by use of an iridium catalyst such as $[Cp^*IrCl_2]_2$, oxidation of the unprotected OH group can be accomplished to obtain the desired product.

Iridium derivatives have been demonstrated to be active catalysts also for oligo and polymerization as well as post-polymerization reactions. Polymerization of terminal alkynes was catalyzed by complexes with phosphines or with nitrogen chelating ligands, but also by simple organoderivatives such as $[Ir(diene)Cl]_2$. Interestingly, by appropriate choice of the non-participative ligands and of reaction conditions it is possible to obtain excellent control of the strereochemistry of the polyene obtained, thus forming either *cis* or *trans*-polyphenylacetylene. At variance, by employing iridium complexes with bidentate phosphines highly regioselective alkyne cyclotrimerization was accomplished, with yields of the asymmetric product 1,2,4-triarylbenzene up to 100%. Finally, the experience gained in transfer hydrogenations on common organic substrates was successfully applied on a post-polymerization reaction, namely the reduction of polyketones to the corresponding polyalcohols: such reaction represents the first example of catalytic reduction of a polyketone.

REFERENCES

[1] Anastas P. T., Kirchhoff M. M. *Acc. Chem. Res.* 2002, 35, 686.

[2] Sheldon R. A. *Chem. Commun.* 2008, 3352.

[3] Crabtree R. H. *Organometallics* 2011, 30, 17 and references cited therein.

[4] Suzuki T. *Chem. Rev.* 2011, 111, 1825 and references cited therein.

[5] see for example (a) Farnetti E., Pesce M., Kaspar J., Spogliarich R., Graziani M. *J. Chem. Soc. Chem. Commun.* 1986, 746; (b) Farnetti E., Nardin G., Graziani M. *J. Chem. Soc. Chem. Commun.* 1989, 1264; (c) Bianchini C., Barbaro P., Scapacci G., Farnetti E., Graziani M. *Organometallics* 1998, 17, 3308.

[6] (a) Jerome F., Pouilloux Y., Barrault J. *Chem. Sus. Chem.* 2008, 1, 586; (b) Behr A., Eilting J., Irawadi K., Leschinski J., Lindner F. *Green Chem.* 2008, 10, 13; (c) Zhou C.-H., Beltramini J. N., FanY.-X., Lu G. Q. *Chem. Soc. Rev.* 2008, 37, 527; (d) Pagliaro M., Ciriminna R., Kimura H., Rossi M., Della Pina C. *Angew. Chem. Int. Ed.* 2007, 46, 4434.

[7] Hekmat D., Bauer R., Neff V. *Process Biochem.* 2007, 42, 71.

[8] Ciriminna R., Palmisano G., Della Pina C., Rossi M., Pagliaro M. *Tetrahedron Lett.* 2006, 47, 6993.

[9] (a) Gallezot P. *Appl. Catal. A* 1995, 133, 179; (b) Dimitralos N., Lopez-Sanchez J. A., Lennon D., Porta F., Prati L., Villa A. *Catal. Lett.* 2006, 108, 147; (c) Carrettin S., McMorn P., Johnston P., Griffin K., Kiely C. J., Hutchings G. J. *Phys. Chem. Chem. Phys.* 2003, 5, 1329; (d) Carrettin S., McMorn P., Johnston P., Griffin K., Kiely C. J., Attard C. A., Hutchings G. J. *Top. Catal.* 2004, 27, 131; (e) Fordham P., Besson M., Gallezot P., *Catal. Lett.* 1997, 46, 195.

[10] (a) Kimura H., Tsuto K., Wakisaka T., Kazumi Y., Inaya Y. *Appl. Catal. A* 1993, 96, 217; (b) Kimura H. *Appl. Catal. A* 1993, 105, 147; (c) Hu W., Knight D.,Lowry B.,Varma A. *Ind. Eng. Chem. Res.* 2010, 49, 10876; (d) Woerz N., Brandner A., Claus P. *J. Phys. Chem.* 2010, 114, 1164.

[11] Painter R. M., Pearson D. M., Waymouth R. M. *Angew. Chem. Int. Ed.* 2010, 49, 9456.

[12] (a) Vinzi, F., Zassinovich, G., Mestroni, G. *J. Mol. Catal.* 1983, 18, 359; (b) Farnetti E., Vinzi F., Mestroni G. *J. Mol. Catal.* 1984, 24, 147.

[13] Farnetti E., Kaspar J., Crotti C. *Green Chem.* 2009, 11, 704.

[14] Gladiali S., Alberico E. *Chem. Soc. Rev.* 2006, 35, 226.

[15] Zassinovich G., Mestroni G., Gladiali S. *Chem. Rev.* 1992, 92, 1051.

[16] (a) Bianchini C., Farnetti E., Graziani M., Nardin G., Vacca A., Zanobini F. *J. Am. Chem. Soc.* 1990, 112, 9190; (b) Farnetti E., Verma Gulati N., Graziani M. *Gazz. Chim. It.*1993, 123, 165.

[17] Crotti C., Kaspar J., Farnetti E. *Green Chem.* 2010, 12, 1295.

[18] Gladiali S., Alberico E., in *Transition Metals for Organic Synthesis*, ed. M. Beller, C. Bolm, 2nd edn., Wiley-VCH, Weinheim, 2004, ch.1.3, pp.145-166.

[19] (a) Fujita K., Tanino N., Yamaguchi R. *Org. Lett.* 2007, 9, 109; (b) van Buijtenen J., Meuldijk J., Vekemans J. A. J. M., Hulshof L. A., Koijman H., Spek A. *Organometallics* 2006, 25, 873; (c) Zhang J., Gandelman M., Shimon L. J. W., Rozenberg H., Milstein D. *Organometallics* 2004, 23, 4026.
[20] Gallezot P. *Green Chem.* 2007, 9, 295.
[21] (a) Sulu M., Venanzi L. M. *Helv. Chim. Acta,* 2001, 84, 898; (b) Jiang Q., Ruegger H., Venanzi L. M. *Inorg. Chim. Acta,* 1999, 290, 64; (c) Barbaro P., Bianchini C., Oberhauser W., A. Togni A. *J. Mol. Catal. A: Chem.*, 1999, 145, 139; (d) Gorla F., Venanzi L. M. *Helv. Chim. Acta,* 1990, 73, 690; (e) Cataldo M., Nieddu E., Gavagnin R., Pinna F., Strukul G. *J. Mol. Catal. A: Chem.*, 1999, 142, 305.
[22] Crotti C., Farnetti E., Guidolin N. *Green Chem.* 2010, 12, 2225.
[23] D'Amato R., Sone T., Tabata M., Sadahiro Y., Russo M. V., Furlani A. *Macromolecules* 1998, 31, 8660.
[24] Yampolskii Yu. P., Korikov A. P., Shantarovich V. P., Nagai K., Freeman B. D., Masuda M., Teraguchi M., wak G. *Macromolecules* 2001, 34, 1788.
[25] Bredas J. L., Adant C., Tackx P. and Persoons A. *Chem. Rev.* 1994, 94, 243.
[26] Weinberge B. R., Ehrenfreund E., Pron A., Heeger A. J., MacDiarmid A. G. *J. Chem. Phys.* 1980, 72, 4749.
[27] Tabata M., Yang W., Yokota K. *J. Polym. Sci. Part A: Polym. Chem.* 1994, 32, 1113.
[28] Kishimoto Y., Eckerle P., Miyatake T., Ikariya T., Noyori R. *J. Am. Chem. Soc.* 1994, 116, 12131.
[29] Kishimoto Y., Eckerle P., Miyatake T., Kainosho M., Ono A., Ikariya T., Noyori R. *J. Am. Chem. Soc.* 1999, 121, 12035.
[30] Misumi Y., Masuda T. *Macromolecules* 1998, 31, 7572.
[31] Kanki K., Misumi Y., Masuda T. *Macromolecules* 1999, 32, 2384.
[32] Katayama H., Yamamura K., Miyaki Y. Ozawa F. *Organometallics* 1997, 16, 4497.
[33] Yao J., Wong W. T., G. Jia G. *J. Organomet. Chem.* 2000, 598, 228.
[34] Hayano S., Masuda T. *Macromolecules* 1998, 31, 3170.
[35] Masuda T., Hayano S., Iwawaki E., Nomura R. *J. Mol. Catal. A: Chem.* 1998, 133, 213.
[36] (a) Vijayaraj T.A., Sundararajan G. *Organometallics* 1997, 16, 4940 ; (b)Bhukta G., Ramanujachary R., Sundararajan G. *J. Organomet. Chem.* 2000, 601, 16.
[37] Mula M. B., Beaumont A. J., Doyle K. O., Gallagher M. L., Rooney A. D. *J. Mol. Catal. A: Chem.* 1999, 148, 23.
[38] Tabata M., Sone T., Sadahiro Y. *Macromol. Chem. Phys.,* 1999, 200, 265.
[39] Falcon M., Farnetti E., Marsich N. *J. Organomet. Chem.*, 2001, 629, 187.
[40] Marigo M., Marsich N., Farnetti E. *J. Mol. Catal. A: Chemical*, 2002, 187, 169.
[41] Marigo M., Millos D., Marsich N., Farnetti E. *J. Mol. Catal. A: Chemical,* 2003, 206, 319.
[42] Fabbian M., Marsich N., Farnetti E. *Inorg. Chim. Acta*, 2004, 357, 2881.
[43] Esteruelas M. A., Olivan M., Oro L. A., Schulz M., Sola E., Werner H. *Organometallics*, 1992, 11, 3659.
[44] Farnetti E., Marsich N. *J. Organomet. Chem.,* 2004, 689, 14.
[45] Ozerov O. V., Patrick B. O., Ladipo F. T. *J. Am. Chem. Soc.* 2000, 122, 6423.
[46] Sugihara T., Wakabayashi A., Nagai Y., Takao H., Imagawa H., Nishizawa M. *Chem. Commun.,* 2002, 576.
[47] Li J., Jiang H., Chen M. *J. Org. Chem.*, 2001, 66, 3627.

[48] (a) Furlani A., Licoccia S., Russo M. V., Camus A., Marsich N. *J. Pol. Sci. Part A: Polymer Chemistry* 1986, 24, 991; (b) Furlani A., Napoletano C., Russo M. V., Camus A., Marsich N. *J. Pol. Sci. Part A: Polymer Chemistry* 1989, 27, 75.

[49] Filipuzzi S., Farnetti E. *J. Mol. Catal. A: Chemical,* 2005, 238, 111.

[50] See for example the following papers and cited references: (a) Ahmetli G., Kaja A., Ozkececi A. *J. Appl. Polym. Sci.*, 2008, 107, 1373; (b) Zhang Y., Broekhuis A. A., Stuart M. C. A., Picchioni F. *J. Appl. Polym. Sci.*, 2008, 107, 262.

[51] Suarez E. J. G., Godard C., Ruiz E., Claver C. *Eur. J. Inorg. Chem.*, 2007, 2582.

[52] Durand J., Milani B. *Coord. Chem. Rev.,* 2006, 250, 542.

[53] Nakano K., Kosaka N., Hiyama T., Nozaki K. *Dalton Trans.* 2003, 4039. Bianchini C., Meli A. *Coord. Chem. Rev.* 2002, 225, 35.

[54] Bianchini C., Meli A. *Coord. Chem. Rev.* 2002, 225, 35.

[55] Jiang Z., A. Sen A. *J. Am. Chem. Soc.*, 1995, 117, 4455.

[56] Nozaki K., Kosaka N., Muguruma S., Hiyama T. *Macromolecules*, 2000, 33, 5340.

[57] Scarel A., Axet M. R., Amoroso F., Ragaini F., Elsevier C. J., Holuigue A., Carfagna C., Mosca L., Milani B. *Organometallics*, 2008, 27, 1486.

[58] Milani B., Crotti C., Farnetti E. *Dalton Trans.*, 2008, 4659.

In: Iridium: Occurrence, Characteristics and Applications ISBN: 978-1- 62081-681-3
Editors: Chieko Fukui and Maemi Ono

Chapter 4

IRIDIUM-CARBONYL COMPLEXES OF NITROGEN-OXYGEN AND PHOSPHROUS-OXYGEN DONOR LIGANDS: SYNTHESIS, CHARACTERISATION AND REACTIVITY

Dipak Kumar Dutta*

Materials Science Division, Council of Scientific and Industrial Research, North-East Institute of Science and Technology, Jorhat, Assam, India

ABSTRACT

Organometallic complexes of iridium have gained much importance because of their novelty, reactivity and applications in diverse fields, for example, the commercial homogeneous catalytic process called *'Cativa'* {Iridium (I) catalyst with Ru-complex activator} for carbonylation of methanol to acetic acid. Iridium carbonyl complexes of nitrogen-oxygen (N-O) and phosphorous-oxygen (P-O) donor ligands are of much interest in recent time. The N-O type ligands (L) like methyl picolinate (*a*), ethyl picolinate (*b*), methyl nicotinate (*c*), ethyl nicotinate (*d*), methyl isonicotinate (*e*) and ethyl isonicotinate (*f*) affords the tetra-coordinated neutral dicarbonyl complexes of the type $[Ir(CO)_2ClL]$. The square planar geometry with Ir in the central position and the intermolecular C–H···O and Ir···C interactions in the complex containing ethyl nicotinate greatly stabilize the supramolecular structure in the solid state. The steric, electronic and kinetics of oxidative addition (OA) of the complexes with electrophiles like CH_3I, C_2H_5I, I_2 etc. are interesting. The complexes, *cis*-$[Ir(CO)_2L_2](PF_6)$ (L = pyridine based ligands) showed Water Gas Shift Reaction (WGSR) activity. The Vaska's complex, *trans*-$[Ir(CO)Cl(PPh_3)_2]$, occupies a central position of pivotal importance in the discovery of OA reactions and several such complexes of ligands, $PMePh_2$, PMe_2Ph, PMe_3 were investigated. The P-O type ligands (L′) like 2-$Ph_2PC_6H_4COOMe$ (*a′*) and 2-$Ph_2PC_6H_4OMe$ (*b′*) afford tetra-coordinated mono-carbonyl Vaska's type complexes *trans*-$[Ir(CO)ClL'_2]$. The IR spectra of such complexes in $CHCl_3$ solution show the terminal υ (CO) bands at around 1957 and 1959 cm^{-1} respectively, which are significantly lower in frequency compared to Vaska's complex, *trans*-$[Ir(CO)Cl(PPh_3)_2]$

* Dipak Kumar Dutta: Fax : +91-376-2370011; Ph. +91-376-2370081; Email: dipakkrdutta @yahoo.com.

(1965 cm^{-1}) and therefore substantiate the enhanced electron density at the metal centre. The molecular structure of the complex containing 2-$Ph_2PC_6H_4COOMe$ ligand indicates iridium–oxygen (ester group) distances [Ir···O (2) 3.24 Å, Ir···O (5) 3.29 Å] and angle [O (5)···Ir··O (2) 157.25°] suggesting a long-range intramolecular "*Secondary*" Ir···O interaction resulting in a *pseudo*-hexacoordinated complex. The complexes undergo OA with small molecules like CH_3I, I_2 and O_2 to produce Ir (III) carbonyl species of the type [Ir(CO)Cl(CH_3)IL$'_2$], [Ir(CO)ClI$_2$L$'_2$] and [Ir(O_2)(CO)Cl(2-$Ph_2PC_6H_4OMe$)$_2$] respectively. Some interesting mixed-ligand complex, [IrCl(CO)(P^iPr_3)(Sb^iPr_3)], was also reported. Other complexes [IrX$(CO)_2$(R_2M)], where, X = Cl, Br, I; M = P, As, Sb; R=C_1-C_{20} alkyl, alkoxy radicals may find application as hydrocarboxylation catalysts. Several water soluble complexes like [Ir(CO)$(TPPMS)_3$]CF_3SO_3, (TPPMS = PPh_2(2-$C_6H_4SO_3K$), and their reactivities with small molecules (H_2 or CO) in polar solvents (DMSO or H_2O) were examined. The complex, [Ir(COD)(PR_3)Py]PF_6, where COD = 1,5-cyclooctadiene, PR_3 = tris-cyclohexylphosphine, and Py = pyridine, known as *Crabtree's* catalyst is active for homogeneous hydrogenation.

INTRODUCTION

Iridium metal atom exhibits [Xe]$4f^{14}5d^76s^2$ electron configuration and can show much wider scope of oxidation states e.g. -3, -1, 0, +1, +2, +3, +4, +5, and +6 valence states in $[Ir(CO)_3]^{-3}$, $[Ir(CO)_3PPh_3]^{-1}$, $[Ir_4(CO)_{12}]^{o}$, [IrCl(CO)$(PPh_3)_2$], [Ir$(mes)_2SEt_2$]$_2$, [$IrCl_3$], $[IrCl_6]^{-2}$, $[Ir(mes)_4]^+$ (mes = mesityl) and [IrF_6] respectively and various coordination geometries in each electron configurations [1-3]. The most common oxidation states are +3 and +4 and iridium complexes normally prefer planar, trigonal-bipyramidal and octahedral structures respectively. Therefore, iridium complexes have great potential for the exploitation of novel catalytic reactions in synthetic organic reactions. A great variety of iridium complexes have been prepared and some of the important routes of preparations are highlighted [Figure 1], where $IrCl_3$. nH_2O is found to react as starting material with different types of ligands [4-9]. Several ligands serve to activate the metal centre in generating coordinatively unsaturated species through opening or liberation of ligands, abstraction of hydrogen, and stabilization of reactive intermediates for catalytic activities. It appears that there are many convenient routes of preparations of iridium metal complexes of varied ligands (*vide infra*) of importance and in particular for catalytic applications.

Platinum metal complexes of unsymmetrical chelating ligands have aroused much interest in recent time because of their structural novelty, reactivity and catalytic applications [10-26] such as hydrogenation, carbonylation and hydroformylation reactions. Homogeneous catalysis by transition metal complexes has become a major synthetic tool in many industrial processes [16]. The potential of such catalytic reactions are due to the fact that most of these reactions are of low energy and selective processes. The pioneer discovery of Wilkinson's catalyst, RhCl $(PPh_3)_3$, during the mid - 1960s, paved the ways for many successful synthesis and applications of rhodium-based catalysts. The catalytic reactions span a wide range of complexes such as from syntheses of fine chemicals like that of *L-DOPA* (a drug for curing *Perkinson's* disease) by an asymmetric hydrogenation of prochiral substrates with rhodium (I)-complexes containing optically active ligands [27] to syntheses of bulk chemicals like carbonylation of methanol for producing acetic acid by *Monsanto's* process [28-30] where $[Rh(CO)_2I_2]^-$ is the active catalytic species (presently owned by B.P. Chemicals, UK). In contrast, breakthrough by iridium complexes are less, for example: Vaska's compound *trans*-[Ir(CO)Cl(PPh$_3$)$_2$], the details of which were first reported in 1965, which can undergo oxidative addition (OA) reactions that serve as key steps in many catalytic reactions. The IrCl(PPh$_3$)$_3$ complex which can be seen as analogous to Wilkinson's catalyst, is not a good hydrogenation catalyst because of the inability of the IrH$_2$Cl(PPh$_3$)$_3$ to release hydrogen. A variety of cationic iridium complexes of phosphines have been used as efficient catalysts for alkene hydrogenation, including the industrial enantioselective hydrogenation of imines. Recently, a new industrial catalytic process called '*Cativa*' has been introduced [31-41] involving iridium complex [Ir(CO)$_2$I$_2$]$^-$, with ruthenium-complex activator for carbonylation of methanol where high efficacy was reported compared to the industrial *Monsanto's* process. The most important catalytic applications of iridium complexes, however, are in the manufacture of fine chemicals, through chemoselective and enantioselective hydrogenation [42].

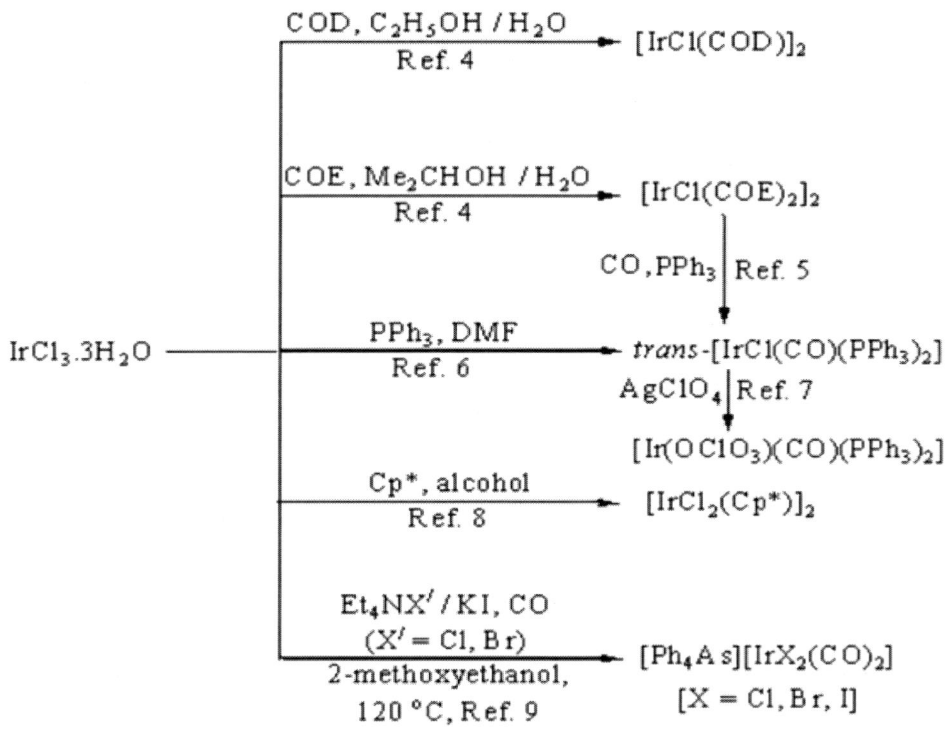

Figure 1. Some important preparative routes for iridium complexes.

IRIDIUM COMPLEXES OF N / N-O DONOR LIGANDS

The coordination chemistry of transition metals in the lower oxidation states have been dominated by phosphorus donor ligands for a long time because of the softness of P-atom due to which they are considered to be more suitable for binding to soft metal atoms [26,43-49]. It has been established recently that many N-donor ligands, in particular *bi*-, *tri*- and tetradentate systems can bind strongly to transition metals stabilizing both high and low oxidation states [50-54]. These ligands are classified based on hybridizations of the N-atom: sp^3, sp^2 and sp. The ligands containing sp^2-hybridized nitrogen atoms, especially when the nitrogen atom is a part of an aromatic system, have extensive coordination chemistry. In contrast to the P-atom, N-atom has only σ-donor and no or very weak π-acceptor [55] properties due to which M-N bond has more prominent ionic character compared to the M-P bond [42,56]. The N-O chelating ligands are also reported, for examples, picolinic acid is known to bind with metal ions as a bidentate N-O donors forming a stable five member chelate ring [57-61]. Interestingly, both the donor sites bear significantly different bonding nature. The N-atom which is a part of benzene ring and the O-atom of the side chain of the benzene ring are "Hard" donors having different hardness stabilize metal centers in their high oxidation states. The combination of atoms of unequal hardness in the same ligand framework may offer the advantage for providing free coordination sites by being detaching the weaker metal-ligand bond reversibly (*hemilabile*) from the coordinating site and make space for the incoming substrate. Thus, N-containing ligands are getting in significance relative to complexes of P-based ligands. Due to strong σ-donor capability, N-donor ligands enhance the nucleophilicity of the metal center which in turn increases the catalytic activity of the complexes. Catalysts that do not have phosphine ligands can offer some advantages as oxidation of phosphines can lead to catalyst deactivation and therefore the catalyst durability may be improved in phosphine-free catalysts [62]. Mono- and bidentate ligands containing N-O donors such as pyridine functionalized with acyl, benzoyl, ester and hydroxyl, have attracted much attention in recent time due to their bonding characteristics. A few iridium (I) complexes of N-O ligands are reported [63-65].

$$[Ir(CO)_2I_2]^- + CH_3I \xrightarrow{OA} \underset{\text{Alkyl intermediate}}{[CH_3Ir(CO)_2I_3]^-} \xrightarrow[\text{CO, Heat}]{\text{Migratory Insertion}} [Ir(CO)(COCH_3)I_3]^-$$

Figure 2. OA reaction of $[Ir(CO)_2I_2]^-$ with CH_3I.

Activation of substrates by transition metal complexes has become most important for both stoichiometric and catalytic reactions. The complexation of substrates by transition metal complexes represent an important step in the course of catalytic processes. The OA is of much importance because it involves scission of comparatively inert substrate bond and thus enable further reactions. The tendency of a metal complex to undergo OA increases with increase in basicity of the metal center [10]. Thus, a low oxidation state in which the metal center is electron rich is favoured and ligands that increase the electron density of the metal center will enhance the reactivity. It is anticipated that a direct interaction of a σ-donor ligand would significantly increase the electron density at the metal atom and consequently lowers the activation barrier of an OA.

The OA of alkyl halide to the metal center is the key rate determining step in the metal complex catalyzed carbonylation of alcohol [66-68]. The oxidative reactivity of the metal complexes depends on the nucleophilicity of the metal center, which in turn depends upon the steric and electronic characteristics of the ligand. The OA of methyl iodide to $[Ir(CO)_2I_2]^-$, on the other hand, yields a stable methyliridium complex, $[MeIr(CO)_2I_3]^-$, for which spontaneous alkyl migration did not occur [49] and only the insertion occurs at elevated temperature in the presence of carbon monoxide gas (Figure 2). Significant weakening of the M-C bonds are required for migration reactions to occur and such bonds to Ir (*5d* elements) are likely to be stronger than to the corresponding Rh (*4d* elements). Therefore, the restricted migratory insertion in iridium species may be due to stronger Ir-C bond. Further, it was also reported [29] that the reactions of $[Ir(CO)_2I_2]^-$ with higher alkyl halides are analogous to the CH_3I, but with much slower rate.

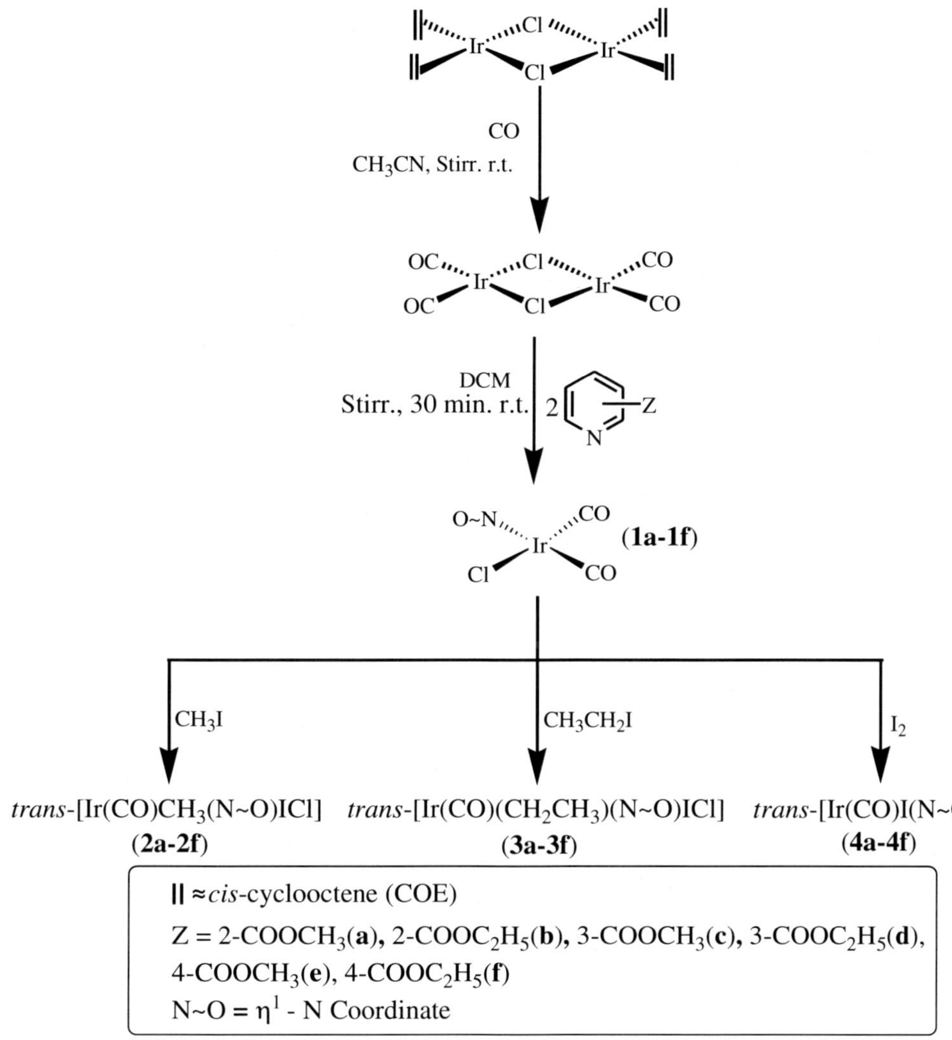

Figure 3. Synthesis and OA reactions of different Ir-complexes.

Recently, Dutta and coworkers reported [69] the reaction of dimeric precursor $[Ir(CO)_2Cl]_2$ with two molar equivalent of the pyridine ester ligands (L) like - methyl picolinate (*a*), methyl nicotinate (*b*), methyl isonicotinate (*c*) ethyl picolinate (*d*), ethyl nicotinate (*e*) and ethyl isonicotinate (*f*) affords the tetra coordinate neutral complexes of the type $[Ir(CO)_2ClL]$ (*1a-1f*). The OA of *1a-1f* with different electrophiles such as CH_3I, C_2H_5I and I_2 undergo decarbonylation of one CO group to generate the oxidized products of the type [Ir(CO)RClIL] where R = $-CH_3$ (*2a-2f*); $-C_2H_5$ (*3a-3f*) and $[Ir(CO)ClI_2L]$ (*4a-4f*) (Figure 3). The single crystal X-ray structure of the complex *1e* reveals that the Ir atom occupies the center of an approximately square planar geometry with two CO groups *cis*- to each other (Figure 4). Intermolecular C-H···O and Ir···C interactions greatly stabilize the supramolecular structure in the solid state (Figure 5). The complex exhibits some interesting intermolecular interactions involving shorter distances than the sum of van der Waals radii. In addition to these hydrogen bonding, a strong intermolecular interaction between iridium metal center and CO group of the adjacent molecule [Ir (1)···C (21) 3.49 (2)] is also observed in their solid state. All these results demonstrate the generation of a supramolecular assembly developed by the extension of a series of helical chains interlinked by C (9)-H (9a)···O (28) and Ir (1)···C (21) intermolecular interactions as shown in Figure 6.

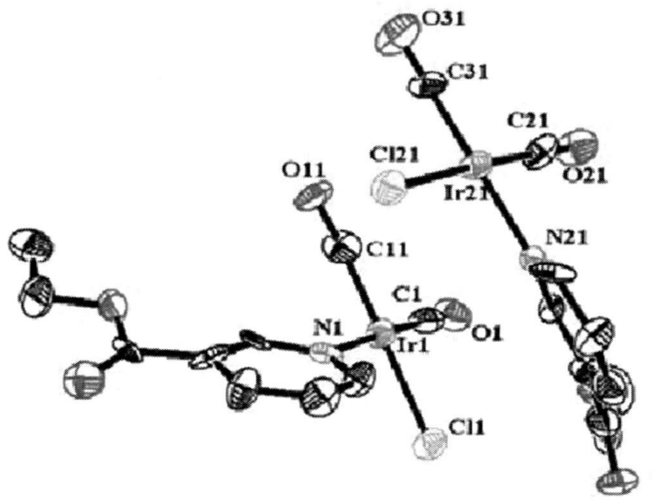

Figure 4. X-ray crystal structure [Ir$(CO)_2$ClL] (L= methyl isonicotinate) (Ref. 69).

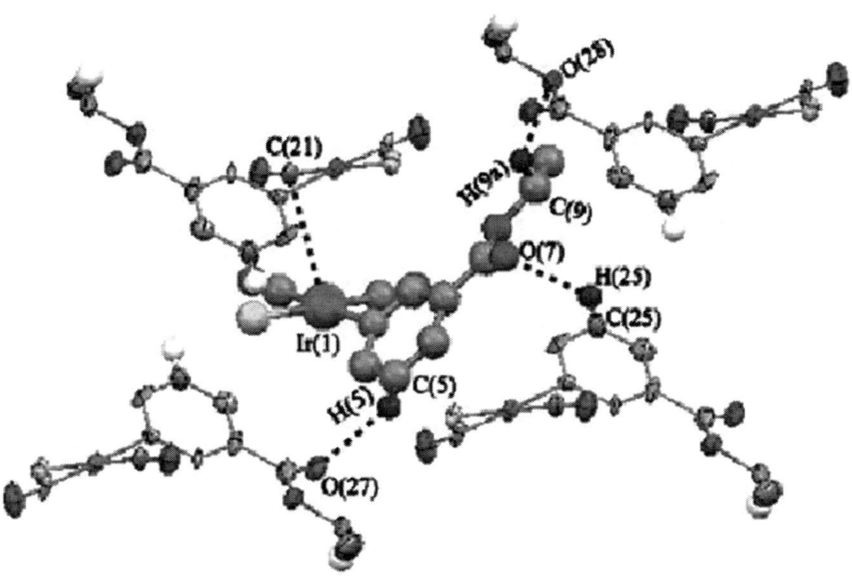

Figure 5. Shortest intermolecular interactions exhibited by mononuclear unit of [Ir$(CO)_2$ClL] (L= methyl isonicotinate) (Ref. 69).

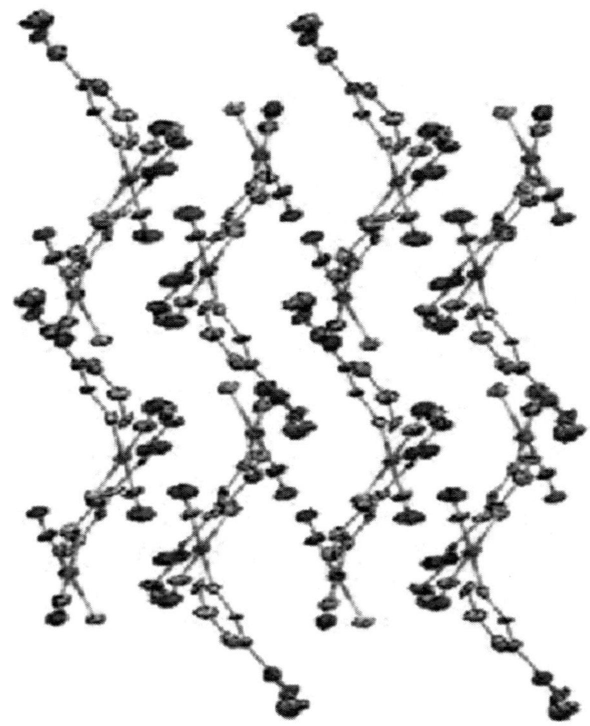

Figure 6. Three dimensional helical structure of $[Ir(CO)_2ClL]$ (L= methyl isonicotinate) (Ref. 69).

The carbonyl-Ir-NHC complexes and the study of their average CO-stretching frequencies [70], have provided information on the electron-donor power of NHC ligands (Figure 7). The high electron-donor power of NHC should favor oxidative addition involving the C-H bonds of their N-substituents (Figure 8).

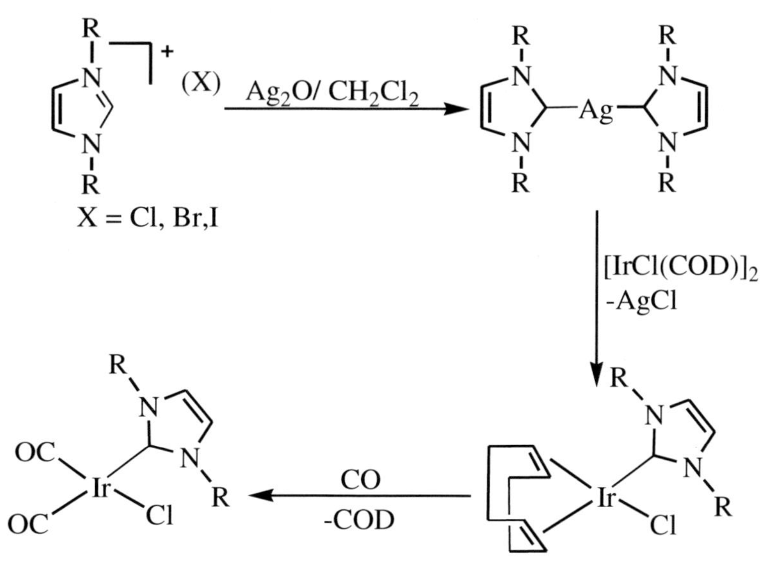

Figure 7. Iridium carbonyl complex.

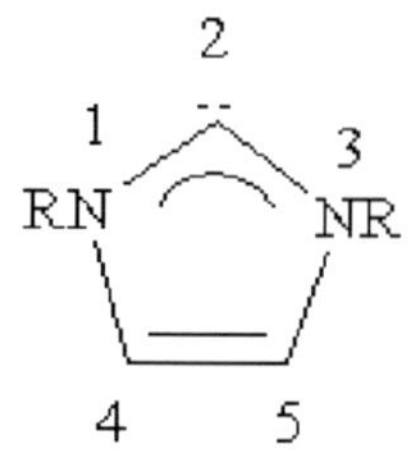

Figure 8. Structure of NHC.

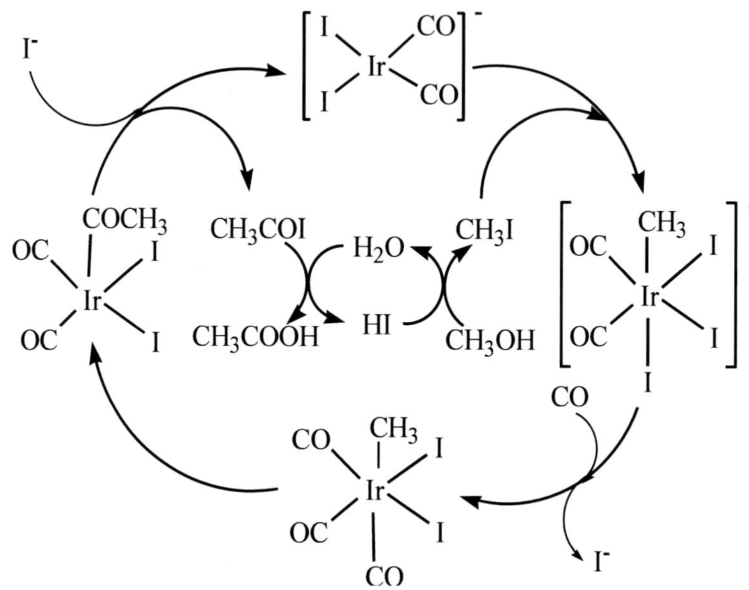

Figure 9. Iridium-based cycle for carbonylation of methanol.

Recently, it was disclosed that an improved iridium catalyst, in combination with a promoter metal such as ruthenium complex, has higher activity and selectivity than reported in previous iridium systems [31-41]. The production of acetic acid using the iridium catalyst system has been commercialized in 1996 by BP Chemicals, UK, as the "*Cativa*" process which is able to operate at reduced water levels (about 5-8 wt. %) than that of 14-15 wt.% for the conventional *Monsanto* process [28-30]. One of the major advantages of the iridium-based process was the high stability of the iridium catalyst species [32]. Besides this stability, iridium is also much more soluble than rhodium in the reaction medium, and thus higher catalyst concentrations can be obtained, making much higher reaction rates achievable. The main difference between the rhodium and iridium-based catalytic cycles has been determined by Maitlis et. al. [71], where the rhodium-based cycle involves only anionic intermediates, but the iridium-based cycle involves both anionic and neutral intermediates. The iridium-based cycle, shown in Figure 9, is similar to the rhodium-based cycle, but operates with different kinetics and the *Cativa* cycle shows advantages over the *Monsanto* process. The OA of methyl iodide to the iridium center is about 150 times faster than the analogous reaction with rhodium [71]. The slowest step in the iridium-based process is the substitution of an iodo ligand by CO. The removal of iodide will promote the rate-limiting step. The presence of a promoter leads to a substantial increase in the proportion of reactive anionic species [(CH_3)Ir

$(CO)_2I_3]^-$ and a considerable decrease in the loss of iridium, by the formation of inactive $[Ir(CO)_3I_3]$ and $[Ir(CO)_2I_4]$-species, which are intermediates in the water-gas shift reaction. Another work [39] presents circumstantial evidence that the main role of the platinum promoter $[PtI_2(CO)_2]$, during the iridium-catalysed carbonylation of methanol is to abstract an iodo ligand from $[IrI_3(CH_3)(CO)_2]^-$ through the formation of the heterodimetallic $[IrI(CH_3)(CO)_2(\mu\text{-}I)_2PtI_2(CO)]^-$ intermediate species.

The iridium complexes, *cis*-$[Ir(CO)_2L_2](PF_6)$ (L = pyridine based ligands) reported by Pardey et.al. [63] showed WGSR activity. The rate follows a first order dependence on P (CO), over the 0.7-1.9 atm. range. The authors proposed a CO addition to the catalyst species prior to rate-limiting step. A mechanistic cycle was proposed. The *cis*-$[Ir(CO)_2L_2](PF_6)$ complex also shows [63] catalytic activities for nitrobenzene reduction to aniline under CO / H_2O. Aniline production (mmol) values follow the order 4-picoline > 2-picoline > 2,6-lutidine > 3-picoline > pyridine > 3,5-lutidine under 1.9 atm of CO at 100 °C; [Ir] = 10 mM, 10 mL of 80 % aqueous amine.

IRIDIUM COMPLEXES OF P / P-O DONOR LIGANDS

Vaska's complex, *trans*-$[Ir(CO)Cl(PPh_3)_2]$, occupies a central position of pivotal importance in the discovery of oxidative addition (OA) reactions of transition-metal complexes [72,73]. The complex with 16 valence electrons, is "Coordinatively unsaturated" one and may bind two-electron ligands to become electronically saturated 18 valence electrons. Upon OA, the oxidation state of the iridium increases from Ir (I) to Ir (III) and the four-coordinated square planar arrangement in the parent complex converts to an octahedral, six-coordinate species. The OA of functionalized alkyl halides to iridium (I) complexes $[IrCl(CO)L_2]$ (L = $PMePh_2$, PMe_2Ph, PMe_3) showed some interesting results [32]. Electrochemical oxidationn of $[IrCl(CO)L_2]$ (L = $PEtPh_2$, PEt_2Ph, and PEt_3) and $[IrX(CO)(PPh_3)]_2$ (X = Cl, Br, I) on a rotating Pt electrode is a diffusion-controlled, irreversible one-electron process, indicating that in the redox addition processes atom transferability is more important than the redox properties of the complex [74]. An interesting complex known as *Crabtree's catalyst* i.e. $[Ir(COD)(PR_3)Py]PF_6$, where COD = 1,5-cyclooctadiene, PR_3 = tris-cyclo-hexylphosphine, and Py = pyridine [75]. It is a homogeneous catalyst for hydrogenation. These compounds are both square planar, d^8 complexes, with a total of 16 valance electrons, which accounts for their reactivity. This catalyst is about 100 times more active than Wilkinson's one and can hydrogenate even *tri-* and tetrasubstituted alkenes. Another series of complexes $[IrX(CO)_2(R_2M)]$, where, X = Cl, Br, I; M = P, As, Sb; R = C_1-C_{20} alkyl, alkoxy radicals were reported as novel complexes which may find application as hydrocarboxylation catalysts, particularly in the absence of added halide promoters [33]. Iridium carbonyl complexes containing two, three and four ligands per iridium are common, and were prepared generally from $IrCl_3.xH_2O$ and the biphyllic ligand under refluxing condition [34] or by dehydrohalogenation of $[IrHX_2(CO)(R_3M)_2]$ with a suitable base like sodium methoxide [35] or interaction of the ligand with $[IrR(CO)_2X_2]^-$ [36]. On the other hand, preparation of iridium carbonyl complexes containing one ligand per one iridium atom was not possible by any of the above processes and any attempt to synthesis such complexes by 1 : 1 ratio of ligand and $[Ir(CO)_2X_2]^-$ reactions, always led to the formation of $[IrHX_2(CO)(R_3M)_2]$ or $[IrHX_2(CO)_2(R_3M)_2]$ indicating that the rate for formation of these complexes were much faster than that of the mono-ligated complexes. Interesting complexes like [Ir(CO)$_3$I], [HIr(CO)$_3$I$_2$], [HIr(CO)$_2$I$_2$(H$_2$O)], [RIr(CO)$_2$I$_2$], where R is an alkyl radical were also reported [37]. The activation of small molecules such as CO, CH$_3$I, H$_2$, O$_2$, etc., by metal complexes is important and interesting [31-37,49,76-81] due to their catalytic reactions like the oxidative addition of CH$_3$I in the rhodium-catalysed *Monsanto* processes [29,30,82] and iridium-based *Cativa* process [31-41,79] process for acetic acid production.

Activation of molecular oxygen by metal complexes has attracted great interest in recent years for chemical processes in terms of cost, atom economy and the potential for limited by-products [79,83-89]. The reversible molecular oxygen carrier [Ir(CO)Cl(PPh$_3$)$_2$] is of great importance [72,73,90], since the 1:1 oxygen adduct [Ir(O$_2$)(CO)Cl(PPh$_3$)$_2$] can be recrystallised and is stable [91-93]. The electronic properties of the ligands for such reactions have got much importance [26,92-97]. The geometrical aspects of the metal environment have also aroused tremendous interest as evidenced by several publications [92,98-102]. The application of complex *trans*-[Ir(CO)Cl(PPh$_3$)$_2$] and others in the hydrogenation of alkenes as well as the polymerization of styrene and carbonylation of methanol to acetic acid are illustrative examples of the applications of Ir (I) complexes in catalytic processes [103-106].

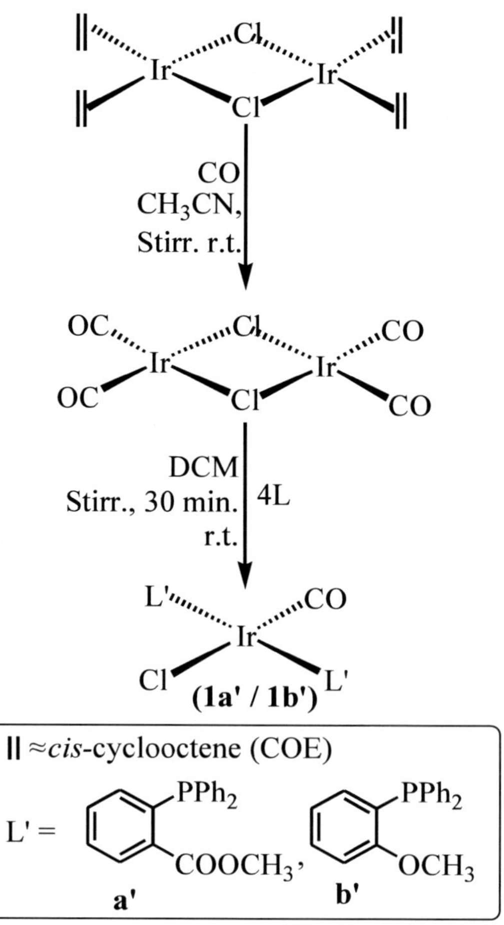

Figure 10. Synthesis of *trans*-[Ir (CO)ClL'$_2$] (L' = 2-Ph$_2$PC$_6$H$_4$COOMe and 2-Ph$_2$PC$_6$H$_4$OMe).

Among the compounds, *trans*-[Ir(CO)Cl(2-$Ph_2PC_6H_4COOMe)_2$] and *trans*-[Ir(CO)Cl(2-$Ph_2PC_6H_4OMe)_2$] [107], the former complex has been found to be inert towards molecular oxygen at room temperature, while the latter is about three times more reactive than Vaska's complex in oxygen uptake, although the physical properties of both complexes closely resemble Vaska's complex. The dimeric precursor [Ir$(CO)_2Cl]_2$ reacts with four molar equivalent of the ligands [L′ = 2-$Ph_2PC_6H_4COOMe$ (*a′*), 2-$Ph_2PC_6H_4OMe$) (*b′*)] by the cleavage of the chloro-bridge to afford tetracoordinated P-bonded complexes of the type *trans*-[Ir (CO)ClL_2] (*1a′, 1b′*) (Figure 10).

The IR spectra of *1a′* and *1b′* in nujol exhibit intense ν (CO) bands at 1940 and 1942 cm^{-1} respectively, indicating the presence of terminal carbonyl groups. These νCO bands occur at significantly lower frequencies than those of the well-known Vaska's complex, (Table 1) and other related Vaska-type complexes [98] like *trans*-[{$(C_6H_4$-2-$CH_3)_3P\}_2$Ir(CO)Cl], *trans*-[{$(C_6H_4$-3-$CH_3)_3P\}_2$Ir(CO)Cl] and *trans*-[{$(C_6H_4$-4-$CH_3)_3P\}_2$Ir(CO)Cl]. The low ν (CO) values of *1a′* and *1b′* may be due to the formation of long-range intramolecular Ir···O secondary interactions by which the ester and ether oxygen atoms donate electron density to the metal centre, which in turn donates electron density to the antibonding π* orbital of CO and consequently lowers the CO bond order. In order to obtain evidence for this intramolecular '*Secondary*' interaction, the IR spectra of *1a′, 1b′* and Vaska's complex were recorded in $CHCl_3$ solution where *1a′* and *1b′* exhibited intense terminal ν (CO) bands at 1957 and 1959 cm^{-1} respectively, both frequencies being significantly lower than that recorded for Vaska's complex, 1965 cm^{-1} (Table 1). The appearance of upfield shifts of CH_3 protons compared to free ligands [δ =3.73 (*a′*) and 3.72 (*b′*) ppm] further substantiate the formation of Ir···O bonds.

Table 1. IR stretching[a] of Vaska's complex, 1a′ and 1b′ and their oxidative addition products

Electrophiles	1a′	1b′	Vaska's complex
None	1940[b]	1942[b]	1952[b]
	1957	1959	1965
CH_3I	2042	2040	2047
I_2	2063	2060	2067
O_2	-	2011	2015

[a] In $CHCl_3$ solution, [b]In Nujol.

The crystal structure of *trans*-[Ir(CO)Cl(2-$Ph_2PC_6H_4COOMe)_2$] (Figure 11) reveals that the iridium atom lies in an approximately square-planar environment formed by the two phosphorus donors, Ir, C (of CO) and Cl atoms. The ester carbonyl oxygen atom of the two phosphane ligands points towards the iridium centre from above and below the vacant axial sites of the planar complex. The spectroscopic evidence (FTIR and ^{1}HNMR spectroscopic data) as well as the observed iridium–oxygen distances [Ir (1)···O (2) 3.246 Å; Ir (1)···O (5) 3.290 Å] (sum of the van der Waals radius of Ir–O ≈ 3.52 Å) indicate long-range intramolecular "*Secondary*" Ir···O interactions leading to a pseudo-hexacoordinated complex. A similar type of long-range interaction was earlier reported by us [26] for a rhodium complex of the 2-$Ph_2PC_6H_4COOMe$ ligand, where the Rh···O distances were 3.08 and 3.18 Å. Another long-range interaction was earlier reported for a palladium complex of 2-phenyl phosphanyl anisole, where the Pd···O distance was 3.17 Å [108]. Miller and Shaw also indicated a typical long-range interaction for the complex trans-[Ir(CO)Cl{PMe_2(2-$MeOC_6H_4$)}$_2$] in which an Ir···O "*Secondary*" bond with the methoxy oxygen was reported [109].

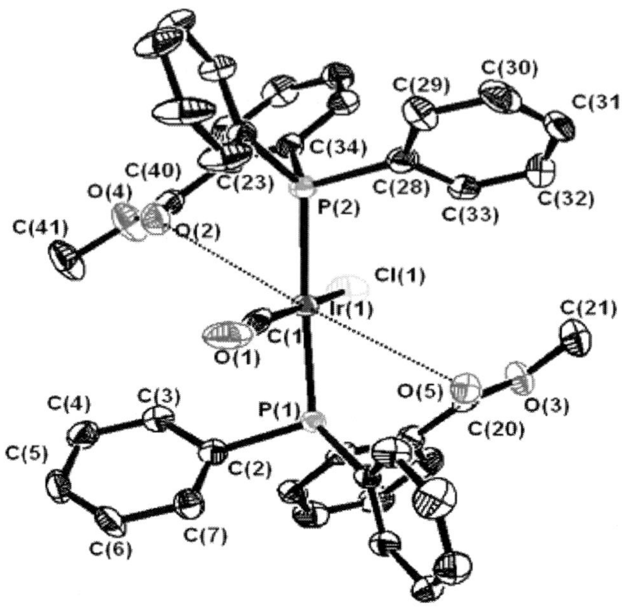

Figure 11. X-ray crystal structure of *trans*-[Ir (CO)Cl(2-$Ph_2PC_6H_4OMe)_2$] (Ref. 107).

An interesting characteristic of square-planar Vaska's complex is that it binds O_2 reversibly through so called '*side-on'* bonding, while in myoglobin and hemoglobin, O_2 binds *'end-on'* to the metal i.e. one of the two oxygen atoms [3]. However, the stereochemical behaviour of the Ir centre is affected by the presence of different coordinating ligands. The complex *trans*-[Ir(CO)Cl(2-$Ph_2PC_6H_4COOMe)_2$] is found to be unreactive towards dioxygen, even after prolonged treatment with O_2. The lack of reactivity of the complex may be due to the steric effect of the ligand 2-$Ph_2PC_6H_4COOMe$ as revealed by the crystal and molecular structure (Figure 11). The two $COOCH_3$ groups, one from each phosphane, are located near the apical positions [(Ir (1)···O (2) 3.246 Å; Ir (1)···O (5) 3.290 Å)] of the planar trans-[Ir (CO)ClP_2] unit, which may effectively protect the central iridium atom from the attacking molecular oxygen.

Such steric inhibition towards dioxygen addition was also reported by Vaska and co-workers [90,98] for the complex *trans*-[{$(C_6H_4$-2-$CH_3)_3P$}$_2$Ir(CO)Cl], in which the steric effect caused by -CH_3 groups was explained with the help of both molecular models and a single-crystal X-ray structure determination. The complex, *trans*-[Ir(CO)Cl(2-$Ph_2PC_6H_4OMe)_2$] on the other hand, was found to be very reactive towards dioxygen and even more reactive than Vaska's complex. The reactivity is increased by strong σ-donor and polarisable ligands, while deactivation is caused by strong π-acceptor ligands, electronegative ligands, a positive charge on the complex. The steric effects of the ligand environment also have a profound influence on the reactivity of the metal centre [98,110,111]. The high reactivity of *trans*-[Ir(CO)Cl(2-$Ph_2PC_6H_4OMe)_2$], may be interpreted by the combined effect of electronic as well as steric factors of the ligands. The higher electron density of the metal centre influenced by the OCH_3 group of the ligand may enhance the susceptibility of the central metal atom towards dioxygen addition. It is expected that during dioxygen addition, the $Ph_2PC_6H_4$-2-OCH_3 ligands may rotate and readjust in such a way that the incoming molecules suffer less steric hindrance as substantiated by Ir···OCH_3 spatial distances [O (5)···Ir (1) 3.546 Å, O (4)···Ir (1) 3.551 Å, C (39)···Ir (1) 4.563 Å, C (20)···Ir (1) 4.409 Å] observed from the X-ray crystal structure of *trans*-[Ir(O_2)(CO)Cl(2-$Ph_2PC_6H_4OMe$)$_2$] (Figure 12).

The complex *trans*-[Ir(CO)Cl(2-$Ph_2PC_6H_4OMe$)$_2$] and Vaska's complex showed addition reactions with O_2 and the rate of reactivities as monitored by UV/Visible spectroscopy exhibit a typical series of spectra (Figure 13) for the reaction of the former complex with O_2 at 25 °C. The bands appearing in the visible (λmax = 434.5, 382.5 nm) and near-visible (λmax = 331.4 nm) region undergo decay until equilibrium is attained. The isosbestic point close to the peak 331.4 nm indicates that the O_2 uptake reaction is proceeding without forming an intermediate or multiple products. A kinetic study of the oxidative addition of methyl iodide to Vaska's complex in ionic liquids was reported [112], where, pseudo-first order rate constants were observed for the oxidative addition reaction in each solvents.

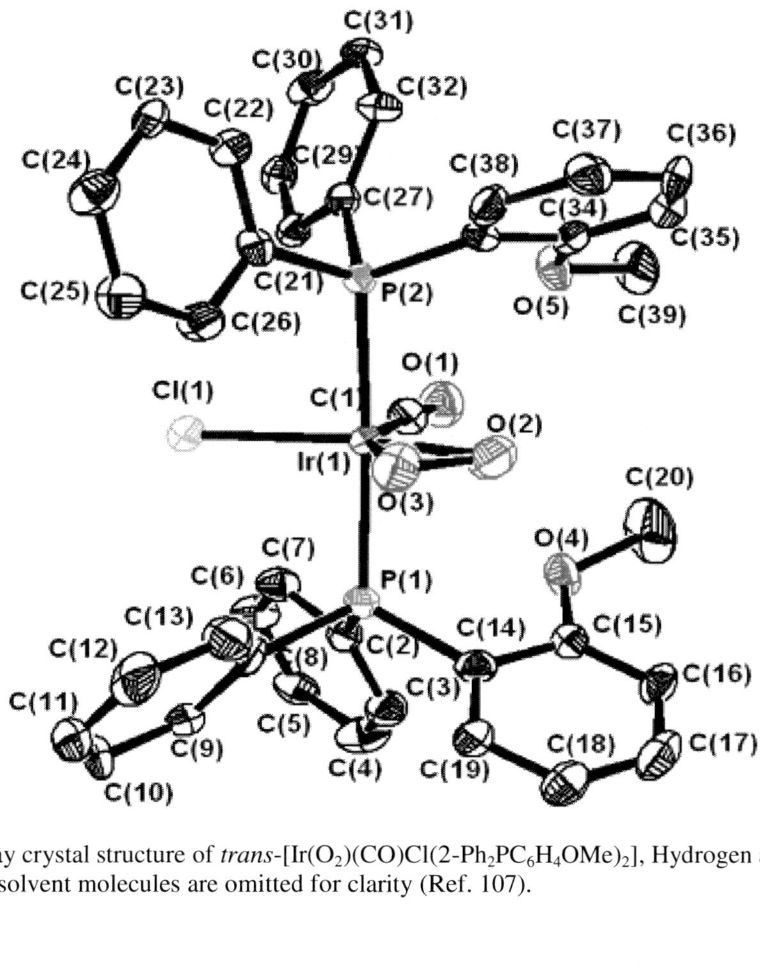

Figure 12. X-ray crystal structure of *trans*-[Ir(O_2)(CO)Cl(2-$Ph_2PC_6H_4OMe$)$_2$], Hydrogen atoms and uncoordinated solvent molecules are omitted for clarity (Ref. 107).

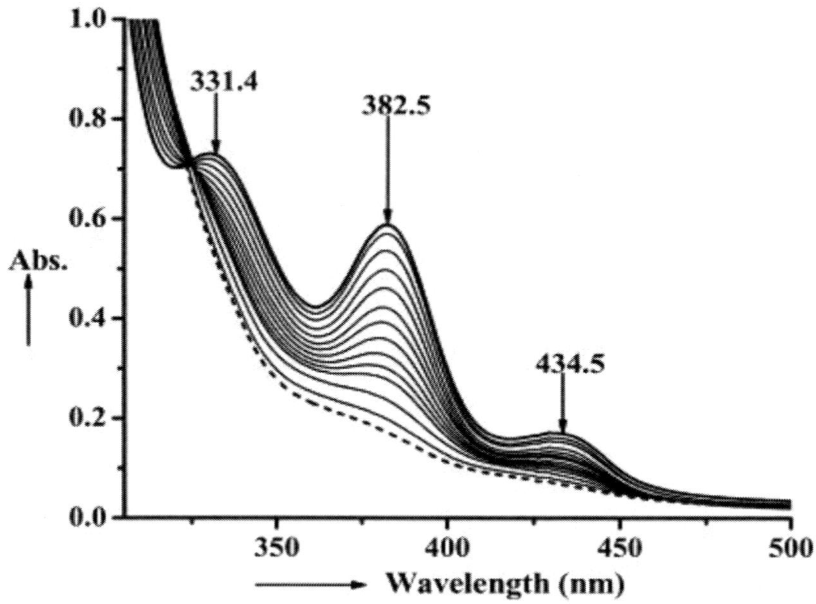

Figure 13. Series of UV/Vis spectra illustrating the reaction of *trans*-[Ir(CO)Cl(2-$Ph_2PC_6H_4OMe)_2$] with molecular O_2 at 25 °C. The arrows indicate the behaviour of each band as the reaction progresses (Ref. 107).

Some interesting mixed-ligand complexes, [IrCl(C_2H_4)(Sb^{i}Pr$_3$)(P^{i}Pr$_3$)] and [IrCl(CO)(P^{i}Pr$_3$)(Sb^{i}Pr$_3$)], were reported [113]. Recently, cationic iridium complexes of the xantphos ligand and their flexible coordination modes and the isolation of the hydride insertion product with an alkene have been reported [114]. Reaction of the Ir (I)-xantphos complex [Ir (η^2-xantphos) (COD)][X] (Xantphos = 4,5-bis (diphenyl-phosphino)-9,9-dimethylxanthene, X = C_6H_3 $(CF_3)_2$) with H_2 in acetone or CH_2Cl_2 /MeCN affords the Ir (III)-hydrido complexes [Ir (η^3-xantphos) $(H)_2$ (L)][X], L = acetone or MeCN, whereas in non-coordinating CH_2Cl_2 solvent dimeric [Ir (η^3-xantphos) (H) (μ-H)]$_2$[X]$_2$ is formed. These complexes show variable coordination modes of the xantphos ligand: *cis*-η^2-P,P, *fac*-η^3-P,O,P and *mer*-η^3-P,O,P.

Effects of cyclopentadienyl and phosphine ligands on the basicities and nucleophilicities of CpIr(CO)(PR_3) complexes were established [115]. The basicities of the series of complexes [CpIr(CO)(PR_3)] {PR_3 = P (4-$C_6H_4CF_3)_3$, P (4-$C_6H_4F)_3$, P (4-$C_6H_4Cl)_3$, PPh_3, P (4-$C_6H_4CH_3)_3$, P (4-$C_6H_4OCH_3)_3$, PPh_2Me, $PPhMe_2$, PMe_3, PEt_3, PCy_3} have been measured by the heat evolved (*Δ*HHM) when the complex is protonated by CF_3SO_3H in 1,2-dichloroethane at 25°C. There is an excellent linear correlation between the basicities (*Δ*HHM) and nucleophilicities (*log k*) of the CpIr(CO)(PR_3) complexes and only the complex CpIr(CO)(PCy_3) with the bulky tricyclohexylphosphine ligand deviates dramatically from the trend. In general, the pentamethylcyclopentadienyl complexes react much faster than the cyclopentadienyl analogs.

Vaska's complex, *trans*-[Ir(CO)Cl($PPh_3)_2$] is known to oxidatively add a broad range of molecules such as H_2, O_2, Cl_2, HCl, CH_3I [72,73,116,117], whereas no example exists for an oxidative addition of a C-C single bond to it [118]. Recently, the oxidative addition of a nonstrained C (sp^2)-C (sp^2) single bond in a Vaska's complex analogue has been reported [118], where, the ligands utilized were P (^{i}Pr)$_2$-O-Ph-CO-Ph-O-P (^{i}Pr)$_2$) (R = H and OMe) contained a benzophenone moiety (Figure 14) and such oxo linkers between the phosphorus groups and the phenyl rings help to provide the shortest possible chelate backbone, forcing and fixing the C-C single bond in close proximity to the metal center. Two *cis*- and *trans*- (in respect of CO and Cl) Vaska type complexes of the ligands were also reported (Figure 15) [118].

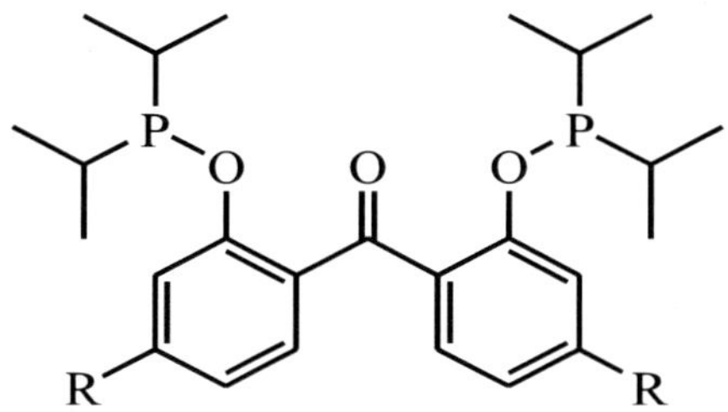

Figure 14. Oxo linkers benzophenone ligand ligand.

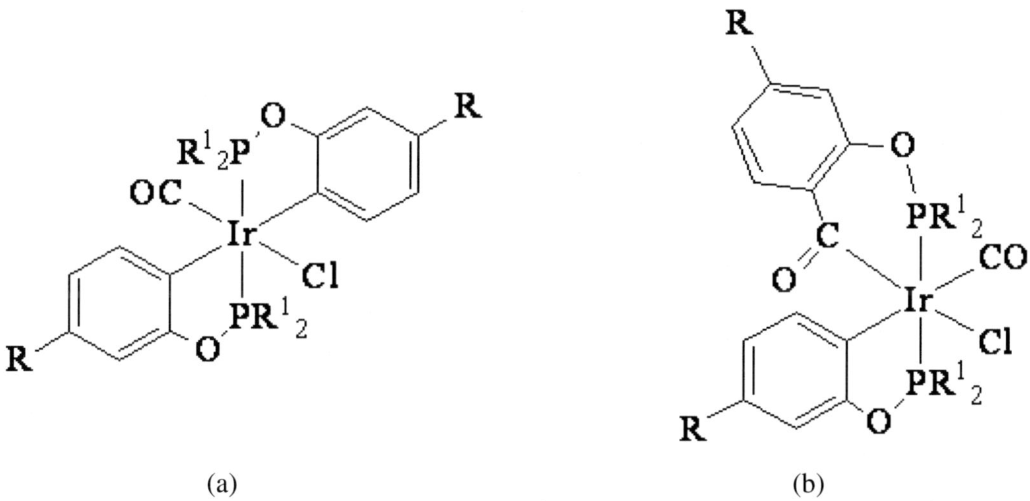

Figure 15. *cis*- (a) and *trans* (b)-Ir-complexes.

Syntheses and characterizations of iridium complexes containing bidentate phosphine ligands and their catalytic hydrogenation reactions to α, β-unsaturated aldehydes were reported [119]. Five iridium complexes containing bidentate phosphine ligands, [IrH(CO)(PPh_3)(BPPB)] [BPPB = 1,2-bis(diphenylphosphino)-benzene], [IrH(CO)(PPh_3)(BISBI)] [BISBI = 2,2'-bis(diphenylphosphinomethyl)-1,1'-biphenyl], [IrH(CO)(PPh_3)(BDNA)] [BDNA = 1,8-bis (diphenylphosphino-methyl)-naphthalene], [IrH(CO)(PPh3)(BDPX)] [BDPX = 1,2-bis(diphenylphosphino-methyl)benzene], and [IrHCl(CO)(PCP)] [PCP-H = 1,3-bis(diphenyl-phosphino-methyl)benzene] were synthesized. The molecular structure of [IrH(CO)(PPh_3)-(BDNA)] determined by single crystal X-ray diffraction indicated a trigonal bipyramidal structure with the three phosphorus atoms in the equatorial plane. The catalytic hydrogenation activities and selectivities of the five complexes as catalysts for citral and cinnamaldehyde were studied [119].

The cationic iridium (I) complexes of water soluble phosphine ligands,

$[Ir(CO)(TPPMS)_3]CF_3SO_3$, $[Ir(CO)(H_2O)(TPPTS)_2]CF_3SO_3$, and $[Ir(CO)_2(TPP\text{-}MS)_3]ClO_4$ (TPPMS = $PPh_2(2\text{-}C_6H_4SO_3K)$, TPPTS = P $(2\text{-}C_6H_4SO_3Na)_3$) were reported [120]. Reaction of H_2 with $[Ir(CO)(TPPMS)_3]CF_3SO_3$ (TPPMS = PPh_2 $(2\text{-}C_6H_4SO_3K)$) in DMSO or H_2O produces [*cis*, *mer*-$Ir(CO)(H)_2(TPPMS)_3]CF_3SO_3$, while the reaction of CO with $[Ir(CO)(TPPMS)_3]CF_3SO_3$ in H_2O yields $[Ir(CO)_2(TPPMS)_3]CF_3SO_3$. Carbonylation of $[Ir(CO)_2(TPPMS)_3]ClO_4$ in DMSO produces $[Ir(CO)_3(TPPMS)_2]ClO_4$ and TPPMS. Hydrogenation of $[Ir(CO)_2(TPPMS)_3]ClO_4$ in DMSO or H_2O yields [*cis*, *mer*-$Ir(CO)(H)_2(TPPMS)_3]ClO_4$, while reaction of H_2 with an aquous solutions of $[Ir(CO)(H_2O)(TPPTS)_2]CF_3SO_3$ produces $[Ir(CO)(H_2O)(H)_2(TPPTS)_2]CF_3SO_3$. Reaction of *trans*-$[Ir(CO)ClL''_2]$ (L'' = TPPMS or TPPTS) with excess L'' in H_2O produces $[Ir(CO)L''_3]Cl$, while no reaction occurs in DMSO. $[Ir(CO)_3(TPPMS)_2]Cl$ reacts irreversibly with TPPMS in H_2O to produce $[Ir(CO)_2(TPPMS)_3]Cl$.

CONCLUSION

Iridium complexes have gained much importance because of their novelty, reactivity and applications in diverse fields. The N-O type ligands (L) like methyl picolinate (*a*), ethyl picolinate (*b*), methyl nicotinate (*c*), ethyl nicotinate (*d*), methyl isonicotinate (*e*) and ethyl isonicotinate (*f*) affords the tetra-coordinated neutral dicarbonyl complexes of the type $[Ir(CO)_2ClL]$. The molecular structure of such complex containing ethyl nicotinate reveals that Ir atom occupies the centre of an approximately square planar geometry and having intermolecular C–H···O and Ir···C interactions greatly stabilize the supramolecular structure in the solid state. The complexes of the type *cis*-$[Ir(CO)_2L_2](PF_6)$ (L = pyridine based ligands) showed Water Gas Shift Reaction (WGSR) activity. Vaska's complex, *trans*-$[Ir(CO)Cl(PPh_3)_2]$, occupies a central position of pivotal importance in the discovery of OA reactions of transition-metal complexes. The P-O type ligands (L′) like $2\text{-}Ph_2PC_6H_4COOMe$ (*a′*) and $2\text{-}Ph_2PC_6H_4OMe$ (*b′*) afford tetracoordinated mono-carbonyl Vaska's type complexes *trans*-$[Ir(CO)ClL'_2]$. The molecular structure of the complex containing $2\text{-}Ph_2PC_6H_4COOMe$ ligand indicates a long-range intramolecular "*Secondary*" Ir···O interaction resulting in a *pseudo*-hexacoordinated complex. The complexes undergo OA with small molecules like CH_3I, I_2 and O_2 to produce Ir (III) carbonyl species of the type $[Ir(CO)Cl(CH_3)IL'_2]$, $[Ir(CO)ClI_2L'_2]$ and $[Ir(O_2)(CO)Cl(2\text{-}Ph_2PC_6H_4OMe)_2]$ respectively. Other complexes [IrX $(CO)_2(R_2M)]$, where, X = Cl, Br, I; M = P, As, Sb; R = C_1-C_{20} alkyl, alkoxy radicals may find application as hydrocarboxylation catalysts. Several water soluble complexes like $[Ir(CO)(TPPMS)_3]CF_3SO_3$, (TPPMS = PPh_2 $(2\text{-}C_6H_4SO_3K)$, and their reactivities with small molecules (H_2 or CO) in polar solvents (DMSO or H_2O) were examined. The complex, $[Ir(COD)(PR_3)Py]PF_6$, where COD = 1,5-cyclooctadiene, PR_3 = tris-cyclohexylphosphine, and Py = pyridine, known as *Crabtree's* catalyst is active for homogeneous hydrogenation. The most important example of application of iridium complexes is the commercial homogeneous catalytic process called *'Cativa'* {Iridium (I) catalyst with Ru-complex activator} developed by BP Chemicals, UK, similar to *Monsanto's* process for carbonylation of methanol to acetic acid.

References

[1] Cotton, F. A.; Wilkinson, G.; Murillo, C.A.; Bochmann, M. *Advanced Inorganic Chemistry*; (6 th ed); John Wiley & Sons: New York, (US), 1999, 1039-1052.

[2] Greenwood, N. N.; Earnshaw, A. *Chemistry of the Elements* (2nd ed.). Oxford: Butterworth–Heinemann. 1997, 1113-1143.

[3] Lee, J. D. *Concise Inorganic Chemistry*, ELBS, (4 th ed.) Chapman & Hall Ltd.: 1991, 783-799.

[4] Herde, J. L.; Lambert, J. C.; Senoff, C. V. *Inorg. Synth.* 1974, 15, 18-20.

[5] Roberto, D.; Cariati, E.; Psaro, R.; Ugo, R. *Organometallics* 1994, 13, 4227-4231.

[6] Vrieze, K.; Collman, J. P.; Sears, C. T.; Kubota, M. *Inorg. Synth.* 1968, 11, 101-104.

[7] Peone, Jr. J.; Vaska, L. *Angew. Chem. Int. Ed.* 1971, 10, 511-512.

[8] Ball, R. G.; Graham, W. A. G.; Heinekey, D. M.; Hoyano, J. K.; Mcmaster, A. D.; Mattson, B. M.; Michel, S. T. *Inorg. Chem.* 1990, 29, 2023-2025.

[9] Forster, D. *Inorg. Nucl. Chem. Letters* 1969, 5, 433-436.

[10] Crabtree, R. H. *The organometallic chemistry of the transition metals*, (2 nd ed.), John Wiely & Sons: New York (US), 1994, 319-325.

[11] James, B. R. *Homogeneous hydrogenation*, John Wiley & Sons: New York, (US), 1973, 288-313.

[12] Cotton, F. A.; Wilkinson, G. *Advanced Inorganic Chemistry, A comprehensive Text*, (4 th ed.), *Interscience,* New York (US), 1980, 1009-1023.

[13] Mullen, A. In *New Synthesis with carbon monoxide;* Falbe, J.; Ed.; Springer–Verlag: Heidelberg, New York, 1980; 11, 275-286.

[14] Dekleva, T. W.; Forster, D. *Advances in catalysis*; Academic Press: 1986; 34, 81-130.

[15] Blaser, H-U.; Indolese, A.; Schnyder, A. *Current Science* 2000, 78, 11, 1336-1344.

[16] van Leeuwen, P. W. N. M. *Homogeneous Catalysis*; Kluwer Academic Publishers: 2004, 112-114.

[17] Cornils, B. *J. Mol. Catal. A* 1999, 143, 1-10.

[18] Thomas, C. M.; Fink, G. S. *Coord. Chem. Rev.* 2003, 243, 125-142.

[19] Howard, M. J.; Jones, M. D.; Roberts, M. S.; Taylor, S. A. *Catal. Today* 1993, 18, 325-354.

[20] Yoneda, N.; Kusano, S.; Yasui, M.; Pujado, P.; Wilcher, S. *Appl. Catal. A* 2001, 221, 253-265.

[21] Bader, A.; Lindner, E. *Coord. Chem. Rev.* 1991, 108, 27-110.

[22] Robertson, R. A. M.; Poole, A. D.; Payne, M. J.; Cole-Hamilton, D. J. *Chem. Commun.* 2001, 47-48.

[23] Rauchfuss, T. B.; Roundhill, D. M. *J. Am. Chem. Soc.* 1974, 96, 3098-3105.

[24] Braunstein, P.; Matt, D.; Dusausoy, Y.; Fischer, J.; Mitschler, A.; Richard, L. *J. Am. Chem. Soc.* 1981, 103, 5115-5125.

[25] Knowles, W. S. *Acc. Chem. Res.* 1983, 16, 106-112.

[26] Dutta, D. K.; Woollins, J. D.; Slawin, A. M. Z.; Konwar, D.; Das, P.; Sharma, M.; Bhattacharyya, P.; Aucott, S. M. *Dalton Trans.* 2003, 2674-2679.

[27] Knowles, W. S. *J. Chem. Educ.* 1986, 63, 222-224.

[28] Paulik, F. E.; Roth, J. F. *Chem. Commun.* 1968, 1578-1582.

[29] Forster, D. *J. Am.Chem. Soc.* 1976, 98, 846-848.

[30] Forster, D. *Adv. Organomet. Chem.* 1979, 17, 255-267.
[31] Sunley, G. J.; Watson, D. J. *Catal. Today* 2000, 58, 293-307.
[32] Bennett, M. A.; Crisp, G. T. *Aust. J. Chem.* 1986, 39, 1363-1375.
[33] Morris, D. E.; Tinker, H. B. *Iridium Carbonyl Complexes*; US Patent no. 948962.
[34] Vaska, L. *J. Am. Chem. Soc.* 1961, 83, 2784-2785.
[35] Deeming, A. J.; Shaw, B. L. *J. Chem. Soc. A* 1968, 1887-1889.
[36] Chatt, J.; Johnson N. P.; Shaw, B. L. *J. Chem. Soc. A* 1967, 604-607.
[37] Forster, D.; Hershman, A.; Morris, D. E. *Iridium Carbonyl Complexes*, US Patent no. 4029748.
[38] Jones, J. *Platinum Metals Rev.* 2000, 44, 94-105.
[39] Gautron, S.; Lassauque, N.; Berre, C. L.; Serp, P.; Azam, A.; Giordano, R.; Laurenczy, G.; Thiébaut, D.; Kalck, P. *Eur. J. Inorg. Chem.* 2006, 1121–1126.
[40] Howard, M. J.; Sunley, G. J.; Poole, A. D.; Watt, R. J.; Sharma, B. K. *Stud. Surf. Sci. Catal.* 1999, 121, 61-68.
[41] Haynes, A.; Maitlis, P. M.; Morris, G. E.; Sunley, G. J.; Adams, H.; Badger, P. W.; Bowers, C. M.; Cook, D. B.; Elliott, P. I. P.; Ghaffar, T.; Green, H.; Griffin, T. R.; Payne, M.; Pearson, J. M.; Taylor, M. J.; Vickers, P. W.; Watt, R. J. *J. Am. Chem. Soc.* 2004, 126, 2847-2861.
[42] Fujita, K.; Yamaguchi, R. In *Iridium Complexes in Organic Synthesis*, Oro, L. A.; Claver, C. Eds., WILEY-VCH: Weinheim, Germany, 2009, 107-140.
[43] Dutta, D. K. In *Organometallic Chemistry: Research Perspective*; Irwin, R. P.; Ed.; Nova Science Publishers Inc. (US): 2007, 249-288.
[44] Das, P.; Sharma, M.; Kumari, N.; Konwar, D.; Dutta, D. K. *Appl. Organomet. Chem.* 2002, 16, 302-306.
[45] Dutta, D. K.; Woollins, J. D.; Slawin, A. M. Z.; Konwar, D.; Sharma, M.; Bhattacharyya, P.; Aucott, S. M. *J. Organomet. Chem.* 2006, 691, 1229-1234.
[46] Gaemers, S.; Sunley, J. G. US Patent no. 7276626 (2007).
[47] Lamb, G.; Clarke, M.; Slawin, A. M. Z.; Williams, B.; Key, L. *Dalton Trans.* 2007, 5582-5589.
[48] Rodriguez, C. J.; Pogorzelec, P. J.; Eastham, G. R.; Slawin, A. M. Z.; Cole- Hamilton, D. J. *Dalton Trans.* 2007, 4160-4168.
[49] Ellis, P. R.; Pearson, J. M.; Haynes, A.; Adams, H.; Bailey, N. A.; Maitlis, P. M. *Organometallics* 1994, 13, 3215-3226.
[50] Haasnoot, J. G. *Coord. Chem. Rev.* 2000, 200-202, 131-185.
[51] Klingele, M. H.; Brooker, S. *Coord. Chem. Rev.* 2003, 241, 119-132.
[52] Kumari, N.; Sarmah, B. J.; Dutta, D. K. *J. Mol. Catal. A* 2007, 266, 260-266.
[53] Sarmah, B. J.; Borah, B. J.; Deb, B.; Dutta, D. K. *J. Mol. Catal. A* 2008, 289, 95-99.
[54] Beckmann, U.; Brooker, S. *Coord. Chem. Rev.* 245, 2003, 17-29.
[55] Lay, P. *Inorg. Chem.* 1984, 23, 4775-4777.
[56] Dutta, D. K.; Chutia, P.; Sarmah, B. J.; Borah, B. J.; Deb, B.; Woollins, J. D. *J. Mol. Catal. A* 2009, 300, 29-35.
[57] Dutta, D. K. *Global J. Inorg. Chem.* 2010, 1, 9-33.
[58] Basu, S.; Peng, S. M.; Lee, G.-H.; Bhattacharya, S. *Polyhedron* 2005, 24, 157-163.
[59] Ugo, R.; Monica, G. L.; Cenini, S.; Bonati, F. *J. Organomet. Chem.* 1968, 11, 159-166.
[60] Barral, M. C.; Aparicio, R. J.; Royer, P. C.; Saucedo, M. J.; Urdanos, F. A.; Pueblo, E. G.; Valero, C. R. *J. Chem. Soc., Dalton Trans.* 1991, 1609-1613.

[61] Ooyama, D.; Kobayashi, T.; Shiren, K.; Tanaka, K. *J. Organomet. Chem.* 2003, 665, 107-113.
[62] Fagan, P. J.; Voges, M. H.; Bullock, R. M. *Organometallics* 2010, 29, 1045-1048.
[63] Pardey, A. J.; Longo, C. *Coord. Chem. Rev.* 2010, 254, 254-272.
[64] Kalck, P.; Serp, P. In *Iridium Complexes in Organic Synthesis*; Oro, L. A.; Claver, C.; Eds.; Wiley-Verlag & Co., Weinheim, 2009, 195-209.
[65] Sengupta, S. K.; Sahni, S. K.; Kapoor, R. N. *Polyhedron* 1983, 2, 317-322.
[66] Haynes, A.; Maitlis, P. M.; Stanbridge, I. A.; Haak, S.; Pearson, J. M.; Adams, H.; Bailey, N. A. *Inorg. Chim. Acta* 2004, 357, 3027-3037.
[67] Das, P.; Konwar, D.; Sengupta, P.; Dutta, D. K. *Trans. Met. Chem.* 2000, 25, 426-429.
[68] Haynes, A.; Mann, B. E.; Morris, G. E.; Maitlis, P. M. *J. Am. Chem. Soc.* 1993, 115, 4093-4100.
[69] Bora, D.; Deb, B.; Fuller, A. L.; Slawin, A. M. Z.; Woollins, J. D.; Dutta, D. K. *Inorg. Chim. Acta* 2010, 363, 1539-1546.
[70] Chianese, A. R.; Li, X. W.; Janzen, M. C.; Faller, J. W.; Crabtree, R. H. *Organometallics* 2003, 22, 1663-1667.
[71] Maitlis, P. M.; Haynes, A.; Sunley, G. J.; Howard, M. J. *J. Chem. Soc., Dalton Trans.* 1996, 2187-2196.
[72] Vaska, L. *Science* 1963, 140, 809–810.
[73] Vaska, L. *Acc. Chem. Res.* 1968, 1, 335–344.
[74] Vecernik, J.; Masek, J.; Vlcek, A. A. *J. Chem. Soc., Chem. Commun.* 1975, 18, 736-737.
[75] Crabtree, R. H.; Morris, G E. *J. Organomet. Chem.* 1977, 135, 395-403.
[76] Griffin, T. R.; Cook, D. B.; Haynes, A.; Pearson, J. M.; Monti, D.; Morris, G. E. *J. Am. Chem. Soc.* 1996, 118, 3029-3030.
[77] Lei, M.; Zhang, W.; Chen, Y.; Tang, Y. *Organometallics* 2010, 29, 543-548.
[78] Adams, R. D.; Captain, B.; Zhu, L. *J. Organomet. Chem.* 2008, 693, 819-833.
[79] Tolman, W. B.; Solomon, E. I. *Inorg. Chem.* 2010, 49, 3555-3556.
[80] Tolman, W. B. *Activation of Small Molecules*; Wiley-VCH: Weinheim, Germany, 2006, 349-350.
[81] Muller, A. J.; Conradie, J.; Purcell, W.; Basson, S. S.; Venter, J. A. *South Afr. J. Chem.* 2010, 63, 11–19.
[82] Forster, D.; Singleton, T. C. *J. Mol. Catal.* 1982, 17, 299-303.
[83] Williams, D. B.; Kaminsky, W.; Mayer, J. M.; Goldberg, K. I. *Chem. Commun.* 2008, 4195-4197
[84] Simandi, L. I. *Dioxygen Activation and Homogeneous Catalytic Oxidation*; Elsevier: New York, 1991.
[85] Rohde, J.-U.; Kelley, M. R.; Lee, W.-T. *Inorg. Chem.* 2008, 47, 11461-11463.
[86] Costas, M.; Llobet, A. *J. Mol. Catal. A* 1999, 142, 113-124.
[87] Shapley, P. A. In *Activation and Functionalization of C–H Bonds*; Goldberg, K. I.; Goldman, A. S., Eds.; chapter 22, ACS Symposium Series 885, 2004, 370- 378.
[88] Limberg, C.; Telesb, J. H. *Adv. Synth. Catal.* 2001, 343, 447-449.
[89] Bakac, A. *Inorg. Chem.* 2010, 49, 3584-3593.
[90] Vaska, L.; Chen, L. S. *J. Chem. Soc., Chem. Commun.* 1971, 1080-1081.
[91] Lee, Y. B.; Wong, W. T. *Chem. Commun.* 2007, 3924-3926.
[92] Vaska, L. *Acc. Chem. Res.* 1976, 9, 175-178.

[93] Lebel, H.; Ladjel, C.; Gariepy, F. B.; Schaper, F. *J. Organomet. Chem.* 2008, 693, 2645-2648.
[94] Roodt, A.; Otto, S.; Steyl, G. *Coord. Chem. Rev.* 2003, 245, 121-137.
[95] Stuurman, N. F.; Conradie, J. *J. Organomet. Chem.* 2009, 694, 259-268.
[96] Johnson, C. E.; Eisenberg, R. *J. Am. Chem. Soc.* 1985, 107, 3148-3160.
[97] Li, X.; Vogel, T.; Incarvito, C. D.; Crabtree, R. H. *Organometallics* 2005, 24, 62-76.
[98] Brady, R.; Camp, W. H. D.; Flynn, B. R.; Schneider, M. L.; Scott, J. D.; Vaska, L.; Werneke, M. F. *Inorg. Chem.* 1975, 14, 2669-2675.
[99] Zinner, S. C.; Rentzsch, C. F.; Herdtweck, E.; Herrmann, W. A.; Kühn, F. E. *Dalton Trans.* 2009, 7055-7062.
[100] Bowden, A. A.; Atwoot, J. D. *J. Coord. Chem.* 1998, 46, 203-209.
[101] Osswald, T.; Ruegger, H.; Mezzeti, A. *Chem. Eur. J.* 2010, 16, 1388-1397.
[102] Perez, S.; Lopez, C.; Bosque, R.; Solans, X.; Bardia, M. F.; Roig, A.; Molins, E.; van Leeuwen, P. W. N. M.; van Strijdonck, G. P. F.; Freixa, Z. *Organometallics* 2008, 27, 4288-4299.
[103] Theron, M.; Grobbelaar, E.; Purcell, W.; Basson, S. S. *Inorg. Chim. Acta*, 2005, 358, 2457-2463.
[104] Gautron, S.; Lassauque, N.; Berre, C. L.; Azam, L.; Giordano, R.; Serp, P.; Laurenczy, G.; Daran, J. C.; Duhayon, C.; Thiebaut, D.; Kalck, P. *Organometallics* 2006, 25, 5894-5905.
[105] Sunley, G. J.; Watson, D. J. *Catal. Today* 2000, 58, 293-307.
[106] Haynes, A.; Meijer, A. J. H. M.; Lyons, J. R.; Adams, H. *Inorg. Chem.* 2009, 48, 28-35.
[107] Dutta, D. K.; Deb, B.; Sarmah, B. J.; Woollins, J. D.; Slawin, A. M. Z.; Fuller, A. L.; Randall, R. A. M. *Eur. J. Inorg. Chem.* 2011, 835-841.
[108] Blin, J.; Braunstein, P.; Fischer, J.; Kickelbick, G.; Knorr, M.; Morise, X.; Wirth, T. *J. Chem. Soc., Dalton Trans.* 1999, 2159-2169.
[109] Miller, E. M.; Shaw, B. L. *J. Chem. Soc., Dalton Trans.* 1974, 480-485.
[110] Valentine, J. S. *Chem. Rev.* 1973, 73, 235-245.
[111] Shaw, B. L.; Stainbank, R. E. *J. Chem. Soc., Dalton Trans.* 1972, 223-228.
[112] P'Pool, S. J.; Klingshirn, M. A.; Rogers, R. D.; Shaughnessy, K. H. *J. Organomet. Chem.* 2005, 690, 3522–3528.
[113] Ortmann, D. A.; Weberndorfer, B.; Ilg, K.; Laubender, M.; Werner, H. *Organometallics* 2002, 21, 2369-2381.
[114] Pontiggia, A. J.; Chaplin, A. B.; Weller, A. S. *J. Organomet. Chem.* 2011, 696, 2870-2876.
[115] Wang, D.; Angelici, R. J. *Inorg. Chem.* 1996, 35, 1321-31.
[116] Vaska, L.; Bath, S. S. J. Am. Chem. Soc. 1966, 88, 1333-1335.
[117] Chock, P. B.; Halpern, J. J. Am. Chem. Soc. 1966, 88, 3511-3514.
[118] Andreas, O.; Klaus, R. *Organometallics* 2011, 30, 171-186.
[119] Li, Rui-Xiang; Li, Xian-Jun; Wong, Ning-Bew; Tin, Kam-Chung; Zhou, Zhong-Yuan; Mak, Thomas C. W. *J. Mol. Catal. A: Chemical* 2002, 178, 181-190.
[120] Paterniti, D. P.; Francisco, L. W.; Atwood, J. D. *Organometallics* 1999, 18, 123-127.

In: Iridium: Occurrence, Characteristics and Applications ISBN: 978-1- 62081-681-3
Editors: Chieko Fukui and Maemi Ono

Chapter 5

PROGRESS IN IRIDIUM-CATALYZED ALLYLIC SUBSTITUTIONS

Hideto Miyabe*
School of Pharmacy, Hyogo University of Health Sciences,
Chuo-ku, Kobe, Japan

ABSTRACT

An important variant for transition metal-catalyzed carbon-carbon or carbon-heteroatom bond formation is allylic substitution. The regio- and stereoselectivities of reactions using iridium catalysts are quite different from those of palladium-catalyzed reactions. Particularly, asymmetric synthesis of branched products based on the regioselective iridium-catalyzed allylic substitution of unsymmetrical substrates has been a subject of current interest. This review highlights the recent studies on iridium-catalyzed allylic alkylation, amination, etherification, and so on. In this review, the enantiocontrol in allylic substitutions using chiral iridium complexes is summarized.

1. INTRODUCTION

Transition metal complex-catalyzed allylic substitutions have been developed as fundamentally important organic reactions [1-9]. The allylic substitutions of symmetrical substrates have been widely studied including enantioselective versions (Scheme 1). Attack by nucleophiles to 1,3-symmetrical π-allyl intermediates *A-1*, generated from an allylic alcohol derivatives *1*, gives allylic products *2*. It has also been important to control the regioselectivity in allylic amination of unsymmetrical substrates. Recently, much effort has been directed toward controlling the regioselectivity as well as enantioselectivity in allylic substitution *via* unsymmetrical π-allyl intermediates *A-2* generated from 1-substituted substrates *3* or 3-substituted substrates *4*. Palladium complexes are general and versatile catalysts for allylic substitution. However, the palladium-catalyzed allylic substitution of *3* or

* E-mail: miyabe@huhs.ac.jp; Fax: +81-78-304-2794.

4 generally gives the achiral linear products *6*. Therefore, the development of catalytic reactions giving the chiral branched products *5* is challenging. The pioneering work has been done by using chiral ferrocenylphosphine-palladium complexes by Hayashi and Ito [10]. Several regio- and enantioselective allylic substitutions using chiral palladium complexes were achieved [11-17].

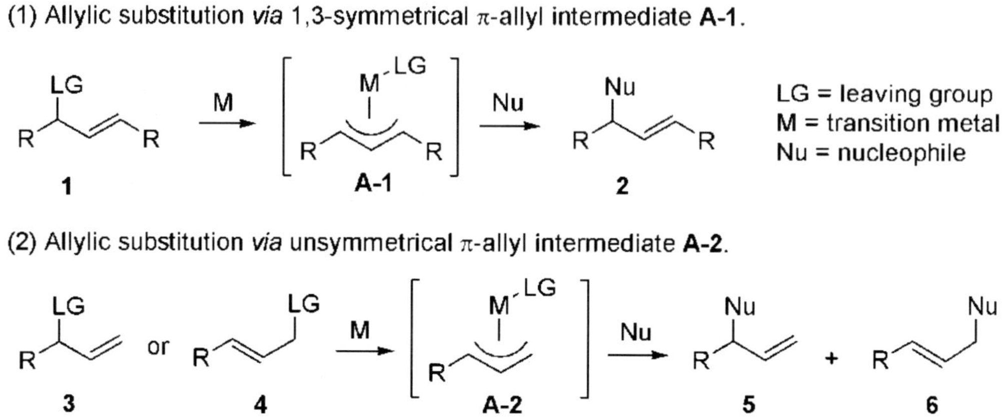

Scheme 1.

Recent studies show that molybdenum, [18-22] rhodium, [23-31] ruthenium, [32-35] tungsten, [36] iron, [37,38] nickel, [39-42] and iridium [43-47] complexes serve as catalysts for allylic substitutions. The regio- and stereoselectivities of reactions using these catalysts are quite different from those of palladium-catalyzed reactions. The allylic substitutions catalyzed by these metal complexes often generate the branched products *5*. Particularly, asymmetric synthesis of branched products *5* by the regio- and enantioselective allylic substitution using chiral iridium complex has been a subject of current interest. This article highlights the recent progress in iridium-catalyzed allylic alkylation, amination, etherification, and so on.

2. Reaction with Carbon Nucleophiles

2.1. Regioselectivity of Iridium-Catalyzed Allylic Alkylation

The regioselectivity of iridium-catalyzed allylic alkylations was initially studied by Takeuchi's group [48,49]. The reaction of (*E*)-2-alkenyl acetate *7* with $NaCH(CO_2Et)_2$ in the presence of a catalytic amount of $[Ir(COD)Cl]_2$ and $P(OPh)_3$ (P/Ir = 2) gave a branched product *8* with 96:4 regioselectivity (Scheme 2). The reaction of regioisomeric acetate *10* gave the same product *8* with high regioselectivity. In these reactions, the phosphorus ligand had a substantial impact on the chemical efficiency and the regioselectivity. The reaction using more electron-donating phosphorus ligands such as PPh_3 gave products in poor yields. With the ligands having weaker π-acceptor property such as $P(OEt)_3$, the reaction was much less selective. Therefore, the π-acceptor property of $P(OPh)_3$ would promote a carbonium ion character at the substituted allylic terminus and directs the nucleophilic attack to this position. The construction of a quaternary carbon center using this methodology was also reported.

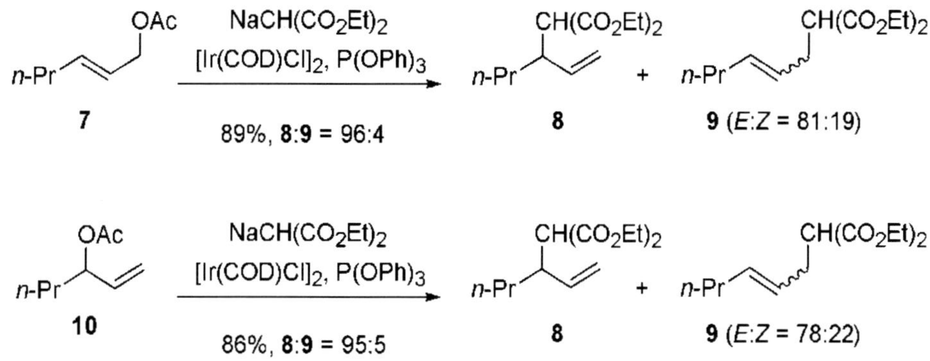

Scheme 2.

2.2. Regio- and Enantioselective Reaction with Malonate

The first regio- and enantioselective allylic alkylation was reported by Helmchen's group [50-56]. The reaction of acetate *11* with $NaCH(CO_2Me)_2$ in the presence of iridium complex of chiral phosphinooxazoline *L1* gave a branched product *12* regioselectively with 95% ee (Scheme 3) [50,51]. The chiral iridium π-allyl complex was characterized by X-ray crystal structure analysis [55]. This study indicates that nucleophile attacks at the central allylic carbon as shown as *B*.

Iridium-catalyzed asymmetric allylic alkylation was investigated by Fuji, Kawabata, and others [57,58]. Chiral aryl phosphite *L2* was used in the reaction of acetate *14* with the anion of dimethyl malonate (Scheme 4). In this study, the enantioselectivities of product *15* were influenced by solvents, countercations, and additives. Particularly, the ee of *15* was remarkably increased to 96% when BuLi and $ZnCl_2$ were employed. Under these conditions, chlorozinc enolate, generated *in situ* from the lithium enolate and $ZnCl_2$, displayed a marked reactivity toward the iridium π-allyl complex. Transition state structure *C* was proposed.

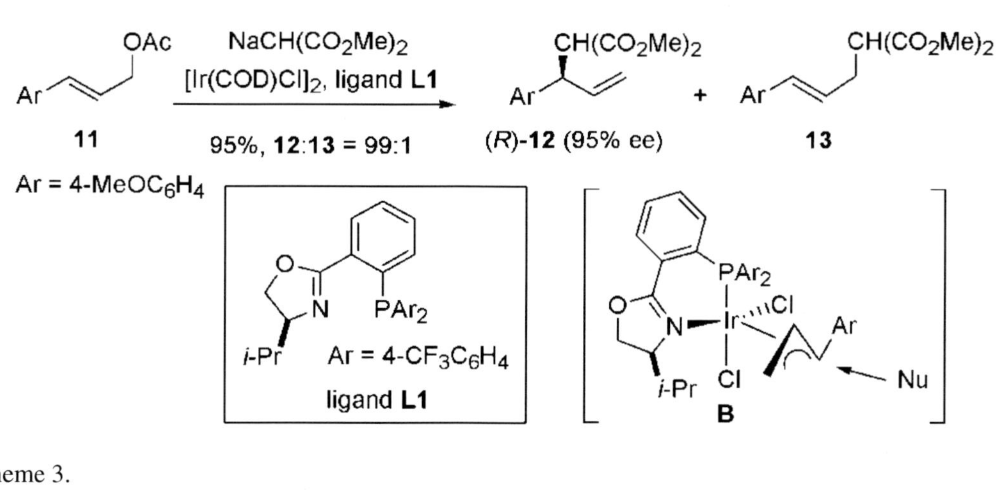

Scheme 3.

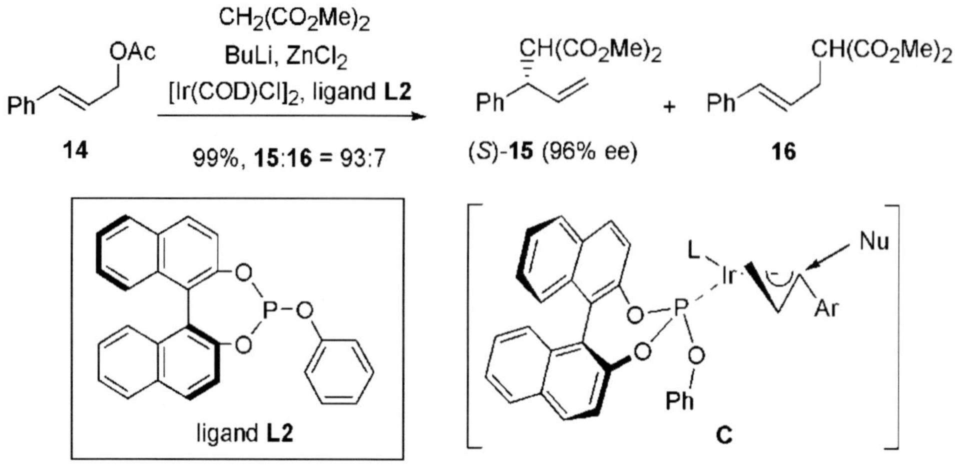

Scheme 4.

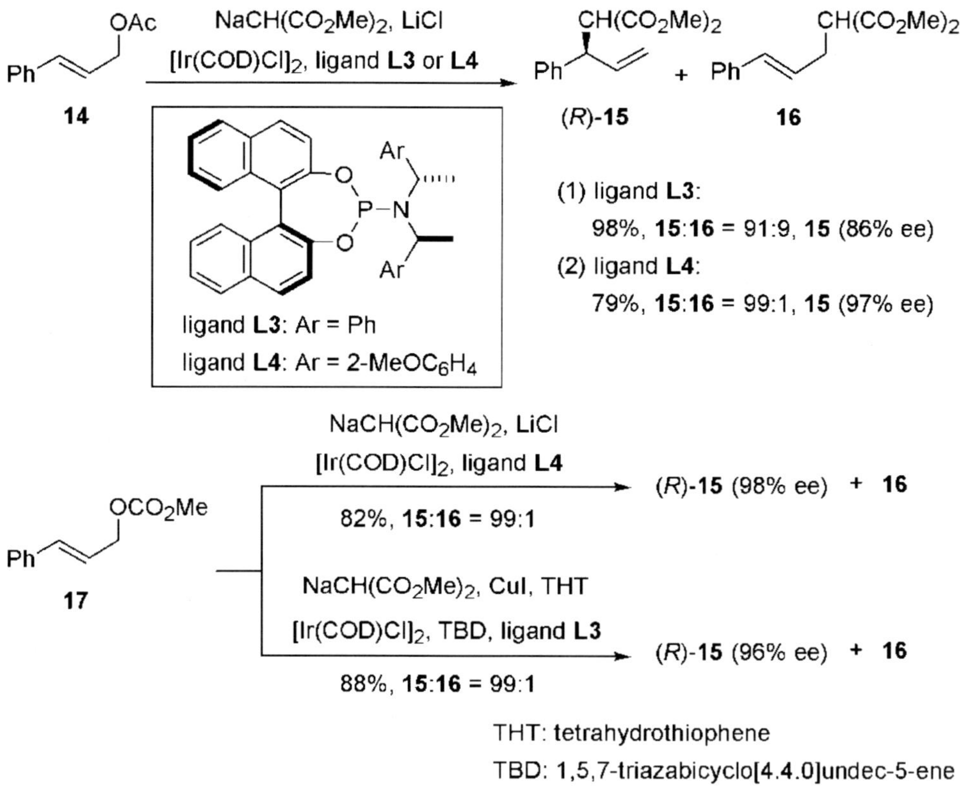

Scheme 5.

The phosphorus amidite ligands, derived from 2,29-dihydroxybinaphthalene (BINOL), are the best ligand for iridium-catalyzed allylic substitutions. The utility of these ligands in allylic alkylation was investigated by Helmchen's group [53,54,56] and Alexakis's group [59-61], respectively. Phosphorus amidite (S_a, *S*, *S*)-*L3*, prepared from bulky secondary chiral amine, induced good enantioselectivities (Scheme 5) [53]. Particularly, the use of LiCl as an additive led to improved regio- and enantioselectivities. For an example, the reaction of acetate *14* gave the branched product *15* with 86% ee. When the ligand *L4* with fixed atropoisomerism was used, enantiomeric excess and regioselectivity were improved to afford the same product *15* with 97% ee [59,61]. Under the similar reaction conditions, the allylic carbonate *17* underwent the enantioselective allylic substitution. Additionally, Helmchen reported that the activity of the chiral catalyst, derived from $[Ir(COD)Cl]_2$ and ligand *L3*, was noticeably increased by using CuI as well as THT as additives in the presence of TBD [54].

2.3. Reaction with other Carbon Nucleophiles

Fluorobis(phenylsulfonyl)methane acts as a synthetic equivalent of CH_2F. You's group reported that the reaction of carbonate *17* with fluorobis(phenylsulfonyl)methane *18* underwent well to afford the branched product *19* with 94% ee (Scheme 6) [62]. The fluorobis(phenylsulfonyl) methylated products could be converted to fluoromethylated compounds without loss of the optical purity.

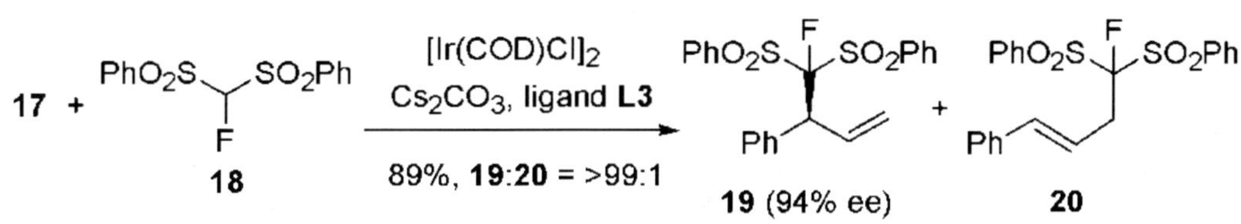

Scheme 6.

The reaction with unsymmetrical nucleophiles is challenging, since regio- and diastereoselectivities as well as enantioselectivity should be controlled [63-65]. The utility of diphenylimino glycinate *22* as an unsymmetrical nucleophile in iridium-catalyzed reactions was investigated by Takemoto, Miyabe, and others (Scheme 7) [63,64]. The bidentate chiral ligand *L5* promoted the reaction of phosphate *21* with *22* with high regioselectivity. The diastereoselective synthesis of both isomers *23* and *24* was achieved by simply switching the base used. When 50% KOH was used, diastereomer *23* was obtained with 97% ee as a major product. In contrast, the use of lithium bases tends to produce another diastereomer *24* as a major product. The stereochemistry can be explained by model *D*. The difference in the behavior of the base might be attributed to the *anti*- and *syn*-geometries of the enolate anions.

2.4. Intramolecular Allylic Alkylation

Intramolecular allylic alkylation of *25* was studied by Helmchen's group (Scheme 8) [56]. Iridium-catalyzed cyclization run after the malonate anion was prepared from *25* by *n*-butyl lithium. A good result was obtained with ligand *L4*. Enantioselective construction of spiroindolenines was investigated You's group [66]. With phosphorus amidite ligand *L6*, the iridium-catalyzed intramolecular C-3 allylic alkylation of indole *27* proceeded smoothly to give the spiroindolenine derivative *28* in 95% yield with 96% ee.

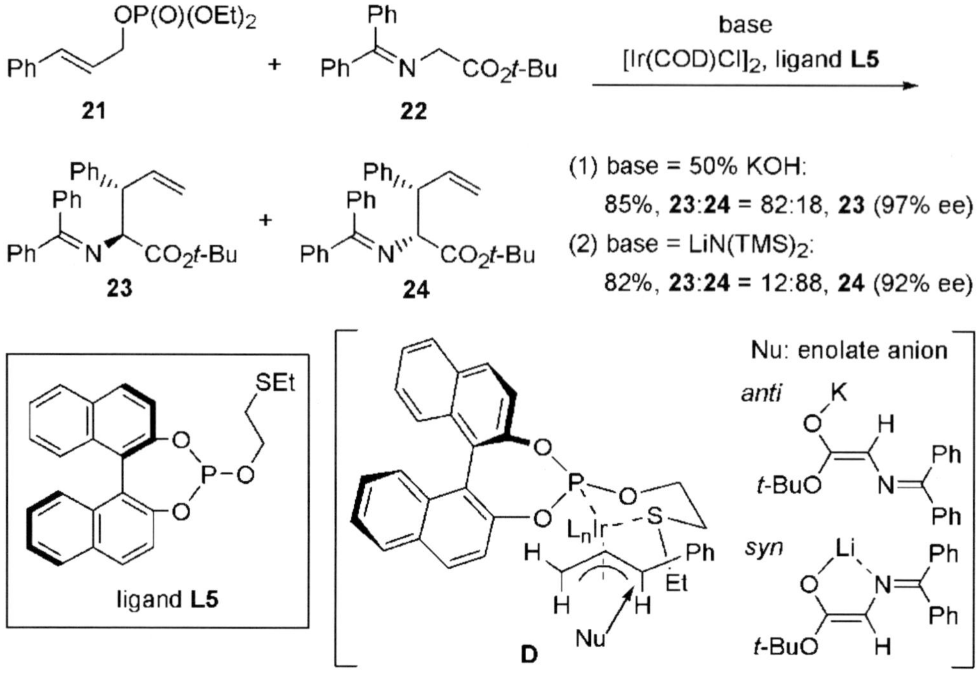

Scheme 7.

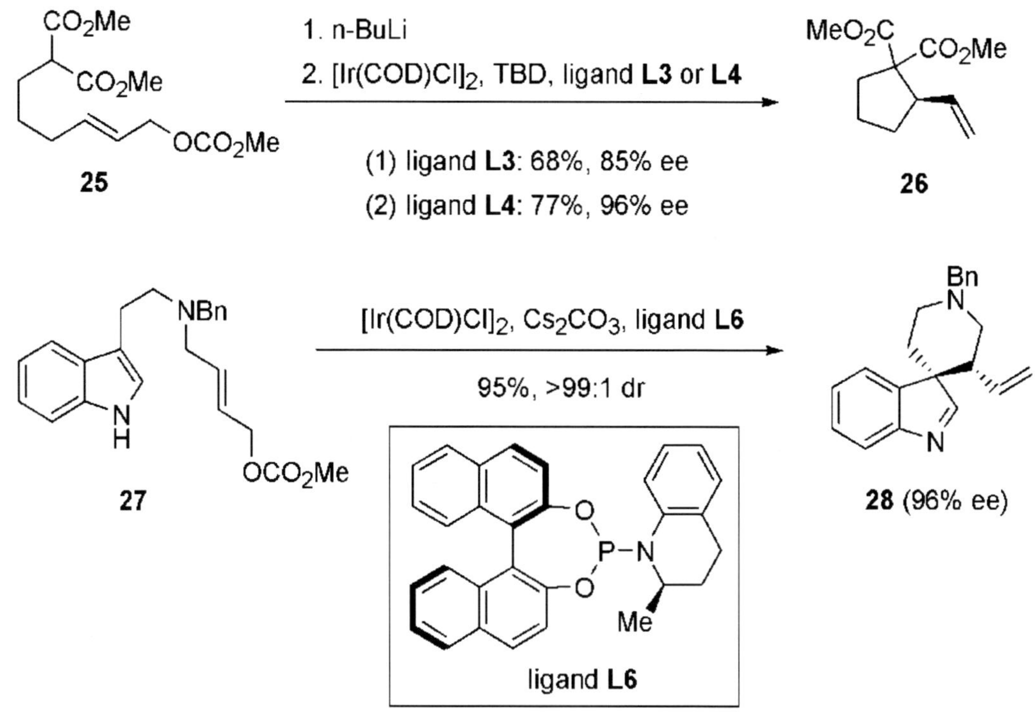

Scheme 8.

3. Reaction with Nitrogen Nucleophiles

3.1. Regioselective Allylic Amination

The regioselective iridium-catalyzed allylic amination was studied by Takeuchi's group [67,68]. The reaction of *17* with piperidine in the presence of $[Ir(COD)Cl]_2$ and $P(OPh)_3$ (P/Ir = 2) gave a branched amine *29* with 95:5 regioselectivity (Scheme 9) [68]. The use of a polar solvent such as EtOH is essential for obtaining the products in high yield. When a primary amine was used under the optimized conditions, selective monoallylation occurred without the formation of diallylation product. In general, the retention of *Z* geometry of substrates in palladium-catalyzed allylic substitution has been difficult to achieve. In the case of iridium-catalyzed allylic amination, the geometry of (*Z*)-*31* was retained to give the linear product (*Z*)-*32* selectively [67,68].

3.2. Regio- and Enantioselective Reaction with Amines

The regio- and enantioselective iridium-catalyzed allylic amination was mainly investigated by Hartwig's group [69-80] and Helmchen's group [81-93]. They found that the iridium complexes of phosphorus amidite ligands catalyzed allylic amination of unsymmetrical substrates with good activities to form the branched products with high enantioselectivities.

Hartwig reported that the iridium complex of phosphorus amidite (R_a, *R*, *R*)-ligand *ent-L3* catalyzed the reaction of carbonate *17* with aliphatic amine such as *n*-hexylamine to give the branched product *34* (R = *n*-Hexyl) with high enantioselectivity (Scheme 10) [69]. Alexakis's group developed that the enantioselectivity could be improved to 98% ee by using the ligand *L4* having 2-methoxyphenyl group [94]. The reaction with aromatic amines was also investigated by using phosphorus amidite ligand *ent-L3* or *ent-L7* [71]. In the presence of ligand *ent-L3* and DABCO (1,4-diazabicyclo [2.2.2]octane) as an activator, the reaction of *17* with aniline proceeded smoothly to give the branched product *34* (R = C_6H_5) with 92% ee. Improvement in enantioselectivity was achieved by conducting reaction with the bulkier ligand *ent-L7*. An activated form of the iridium complex was identified by using ligand (S_a, S, *S*)-*L3* [70,72]. The activated complex was formed by *in situ* C-H activation at CH_3 group of a hindered ligand *L3*. The use of this cyclometalation species as a catalyst led to improved activity, selectivity and broader substrate scope for allylic aminations [73,77,79,80]. The direct iridium -catalyzed amination of allylic alcohols was also studied in the presence of Lewis acid activators. [76]

The reaction of dienyl carbonates *36* using the ligand *L3* was investigated by Helmchen's group (Scheme 11) [81]. High regioselectivities in favor of the desired internal substitution products *37* were obtained.[9a] The aminations provided ee values up to 97%. An activate iridium complex of ligand *L3* and its reaction mechanism were also studied [91,92]. Additionally, several nitrogen nucleophiles having the removal protecting groups on nitrogen atom were tested by using the ligand *L3*. [84,86]

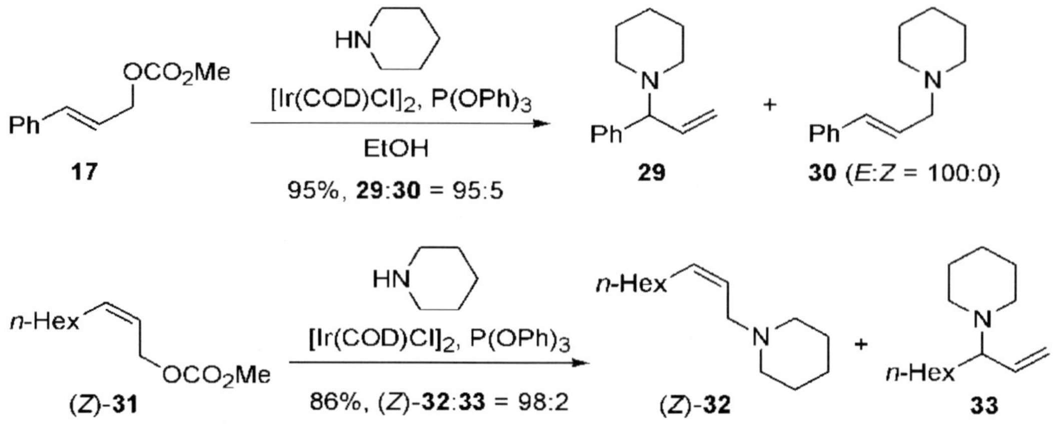

Scheme 9.

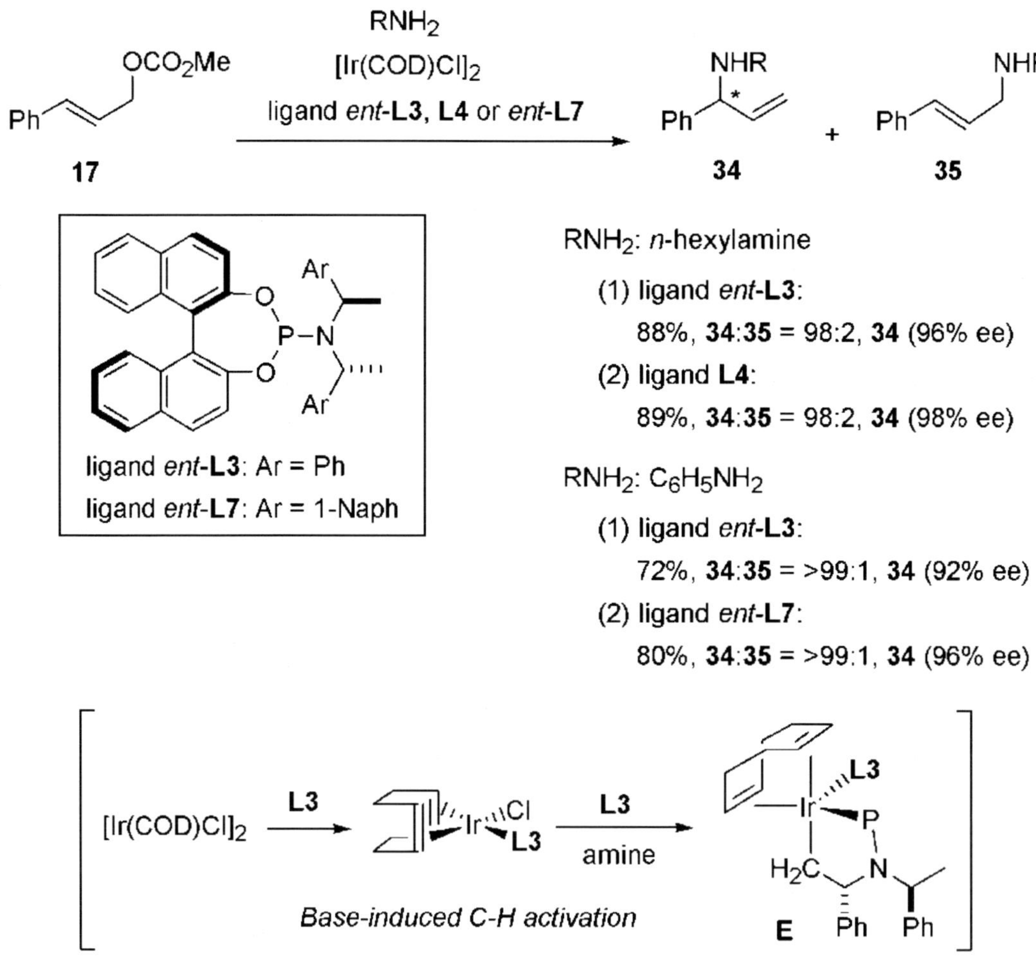

Scheme 10.

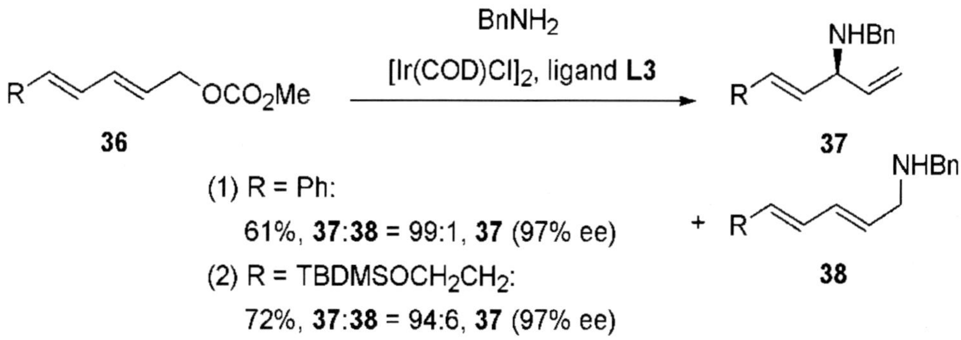

Scheme 11.

Takemoto, Miyabe, and others reported that the iridium-complex of pybox ligand *L8* catalyzed the allylic amination to form the branched products with good enantioselectivities (Scheme 12) [95-97]. In the presence of $CsOH{\cdot}H_2O$, the reaction of phosphate *21* with BnONHBz *39* proceeded smoothly to give the branched product *40* with 92% ee. [95,97]

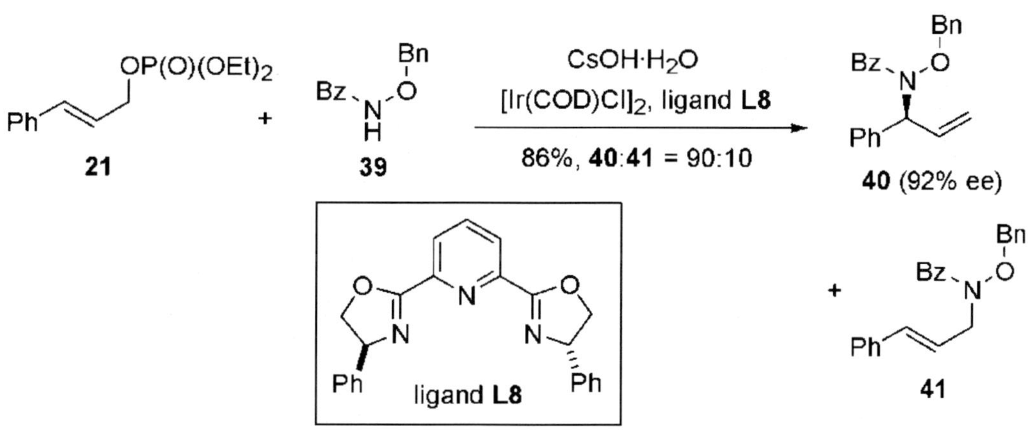

Scheme 12.

The iridium-catalyzed allylic amination using chiral diaminophosphine oxide ligand *L9* was reported by Hamada's group (Scheme 13) [98]. In the presence of BSA and $NaPF_6$, the reaction of *17* with benzylamine afforded the chiral branched allylic amine *42* with 92% ee. In this reaction, the diaminophosphine oxide ligand *L9* is activated by BSA-induced tautomerization to afford trivalent phosphorus compound, which functions as the actual ligand. Additionally, this amination was accelerated by the addition of $NaPF_6$.

Carreira's group reported the direct iridium-catalyzed synthesis of primary allylic amines from allylic alcohols (Scheme 14) [99]. In this reaction, sulfamic acid *45* serves not only as a nitrogen source but also as an *in situ* activator of hydroxy group of allylic alcohols through the condensation with DMF (*N,N*-dimethylformamide). The allylic amination of allylic alcohol *44* using ligand *L10* occurred with high regioselectivity to give (*S*)-1-cyclohexylprop-2-en-1-amine hydrochloride *46* with 70% ee.

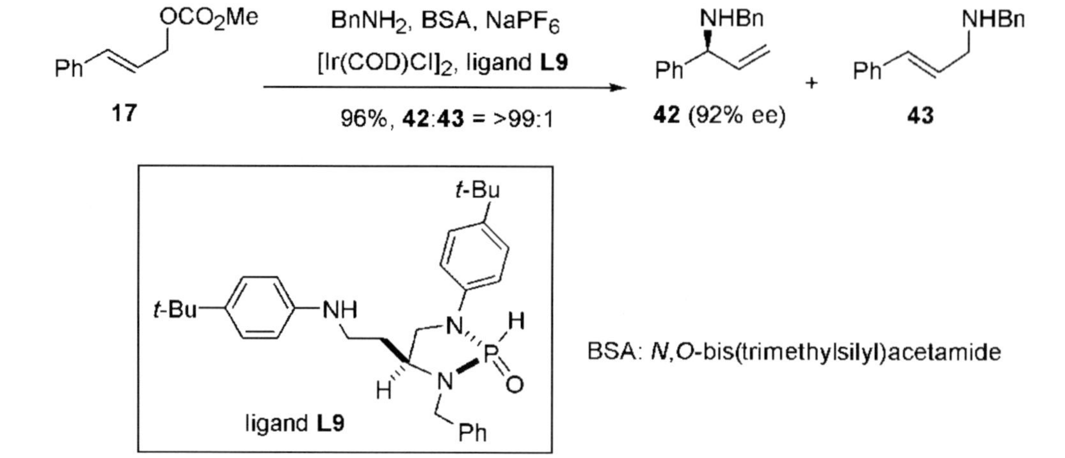

Scheme 13.

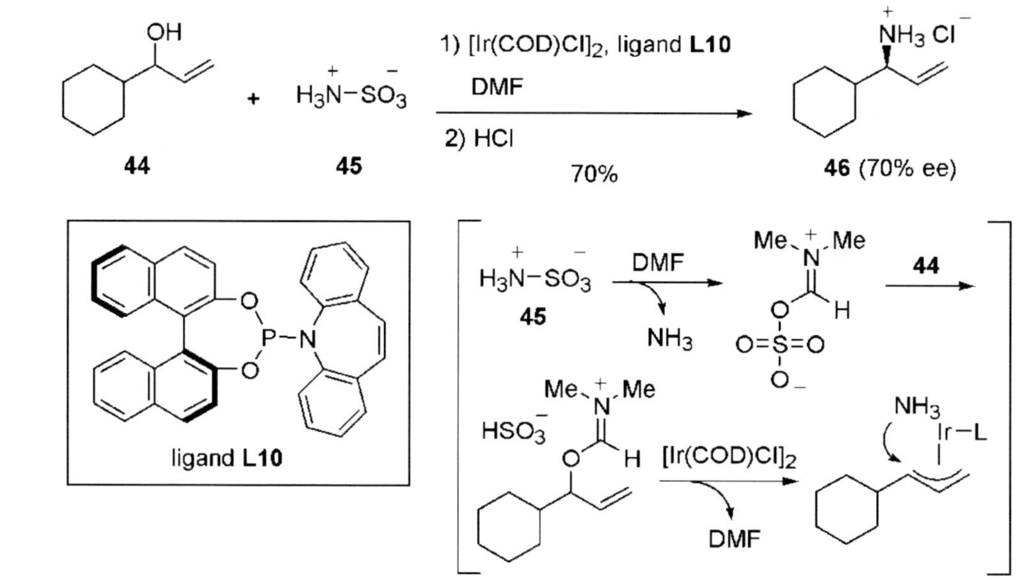

Scheme 14.

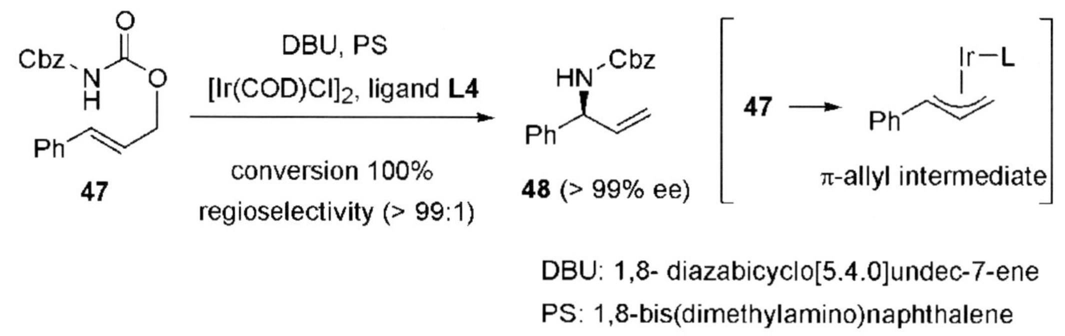

Scheme 15.

The decarboxylative allylic amidation reaction of imidodicarbonates was studied by Han's group (Scheme 15) [100-102]. The iridium-catalyzed reaction of cinnamyl benzyl imidodicarbonate *47* using ligand *L4* provided easy access to chiral product *48* with excellent regio- and enantioselectivities [100]. Both DBU and PS (proton sponge) were necessary for this transformation, since the reaction did not proceed to completion in the absence of either. The result of a crossover experiment was more consistent with the reaction mechanisms involving the π-allylic intermediate over intramolecular [3,3]-rearrangement.

3.3. Reaction with other Nitrogen Nucleophiles

The iridium-catalyzed reaction with the heterocyclic nitrogen nucleophiles was investigated by Hartwig's group [78]. In this study, the single component, ethylene-bound, metallacyclic iridium catalyst *C1* was used (Scheme 16). In the presence of K_3PO_4 and this catalyst *C1*, the reaction of *17* with benzimidazole *49* gave the *N*-allylated benzimidazole *50* with good regio- and enantioselectivities. Simple imidazoles and purines also acted as effective azole nucleophiles.

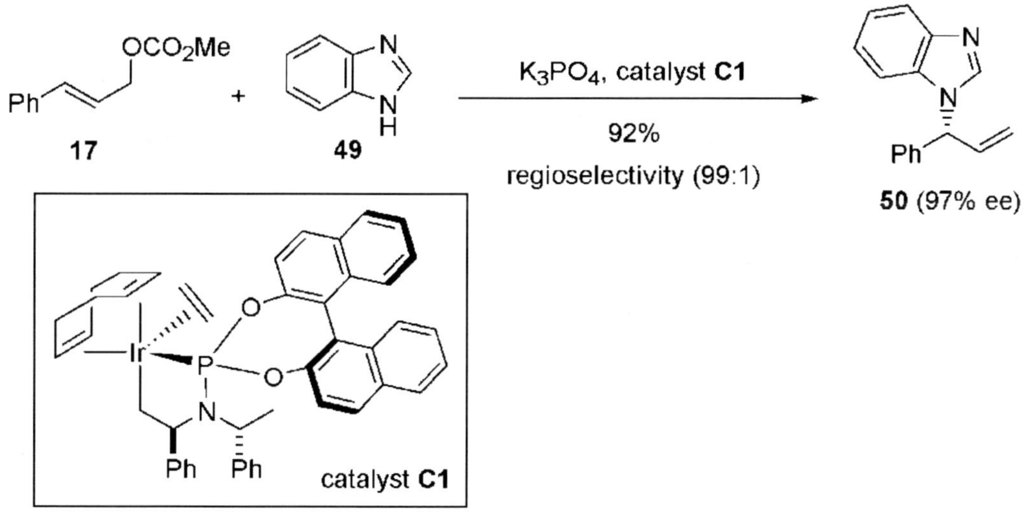

Scheme 16.

As a convenient and direct functionalization of guanidines, the transition metal-catalyzed allylic substitution of guanidines was studied by Takemoto and Miyabe [96]. The guanidine derivative *52* bearing two electron-withdrawing substituents acted as a reactive nucleophile in the allylic substitution of *51* using the iridium complex of chiral pybox ligand *L8* to give the monoallylated product *53* with 96% ee (Scheme 17). The double allylic substitution was achieved by using tri-Boc-guanidine *55* bearing three electron-withdrawing substituents to give the diallylated product *56* with high diastereo-, regio- and enantioselectivities *via* the kinetic resolution on the second allylation step of monoallylated product. The stereochemical feature can be rationalized in terms of steric control in π-allyl iridium complex *F*.

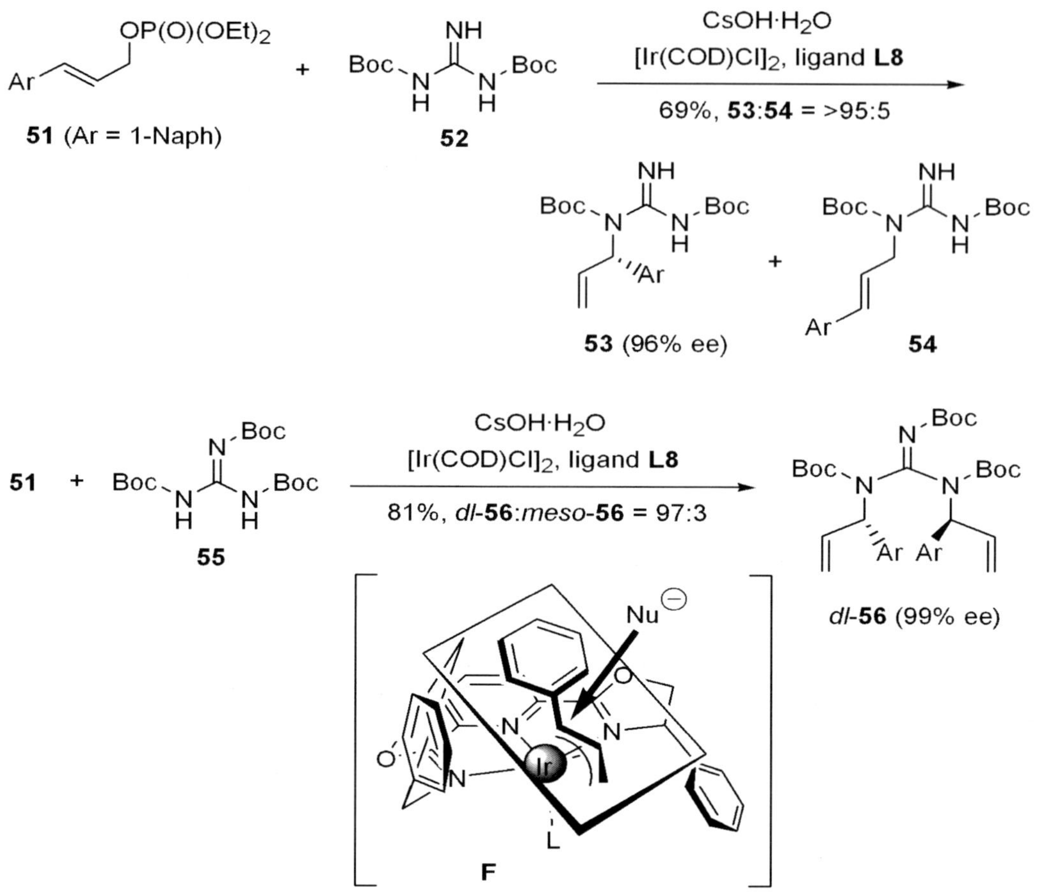

Scheme 17.

3.4. Iridium-Catalyzed Cyclization

Allylic amination is important for the synthesis of nitrogen-containing heterocycles. Takemoto and Miyabe reported a novel method for preparing azacycles *58* based on the sequential allylic amination of *57* (Scheme 18) [103]. The success of this method reflects a high degree of regiocontrol in inter- and intramolecular iridium-catalyzed allylic aminations. Both nitrogen and oxygen atoms on hydroxylamines with an *N*-electron-withdrawing substituent (EWG-NHOH) acted as nucleophiles. Thus, the cyclization of *57* using EWG-NHOH gave the six-membered products *59*.

Helmchen's group has investigated the enantioselective intermolecular reactions [82,83,84,89,90]. The cyclization reaction of *60* using the iridium complex with ligand *L3* was reported (Scheme 19) [82]. The iridium-*L3* complex was treated with a variety of bases, because activated iridium complex is generated by C-H activation promoted by the nucleophile or base. The best enantioselectivity was obtained upon activation with TBD as base to give the cyclic product *61* with 92% ee. Sequential aminations of bis-allylic carbonate *62*, involving an inter- followed by an intramolecular reaction, gave *trans-63* with >99% ee [83].

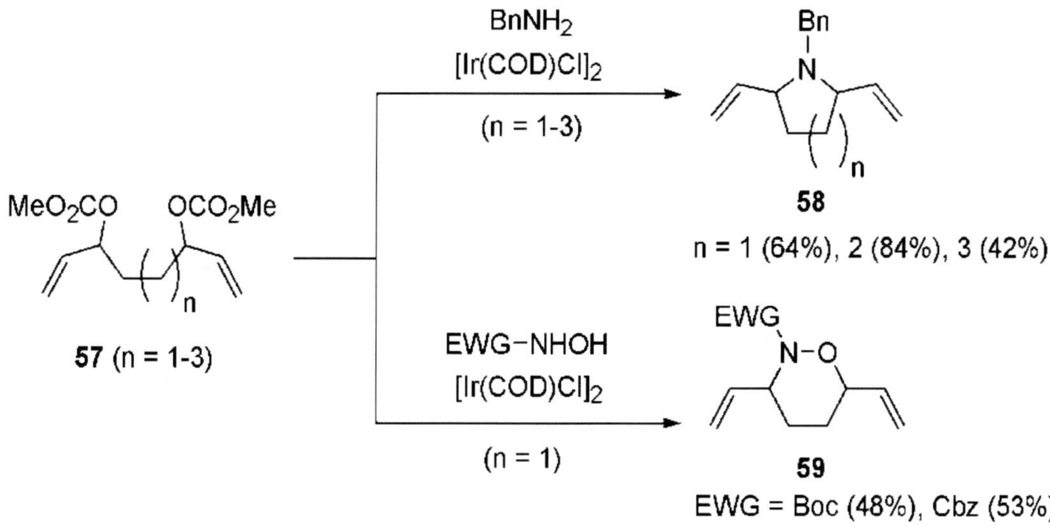

Scheme 18.

Iridium-catalyzed intramolecular allylic amidation was studied for the synthesis of tetrahydroisoquinolines by Feringa's group [104]. The conversion of allylic carbonate *64* into chiral tetrahydroisoquinoline *65* proceeded smoothly with good regio- and enantioselectivities by using ligand *L4* (Scheme 20). You's group studied the enantioselective synthesis nitrogen heterocycles using iridium-catalyzed reactions [105,106]. He reported the new tandem allylic vinylation and intramolecular allylic amination reaction of dicarbonates based on their study on iridium-catalyzed allylic vinylation reaction [105]. In the presence of DABCO and ligand *L3*, 2-vinylaniline *67* reacted with dimethyl dicarbonate *66* to give the seven-membered product *68* in 73% yield with 92%ee.

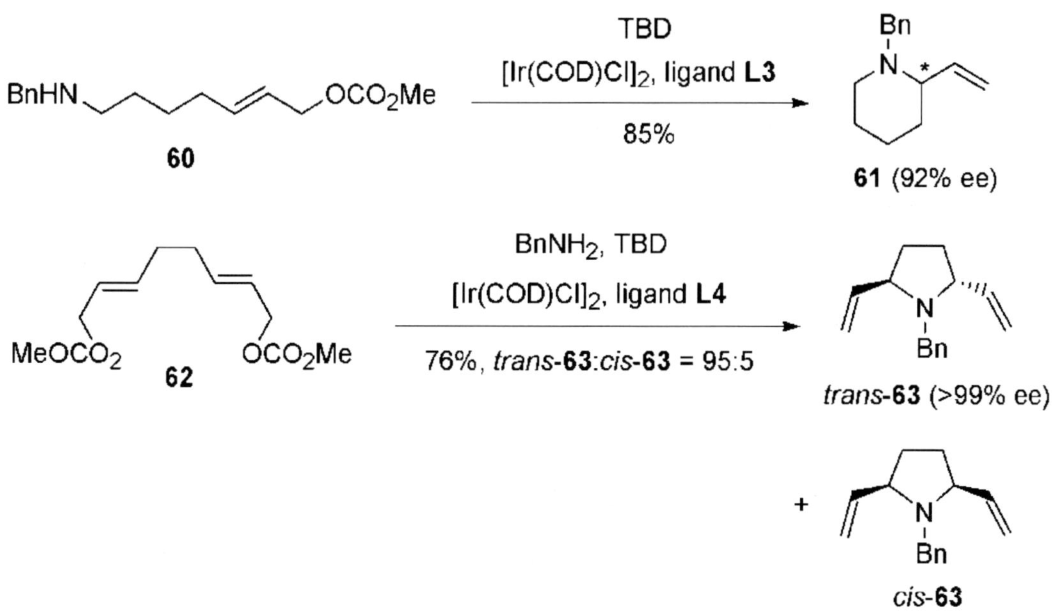

Scheme 19.

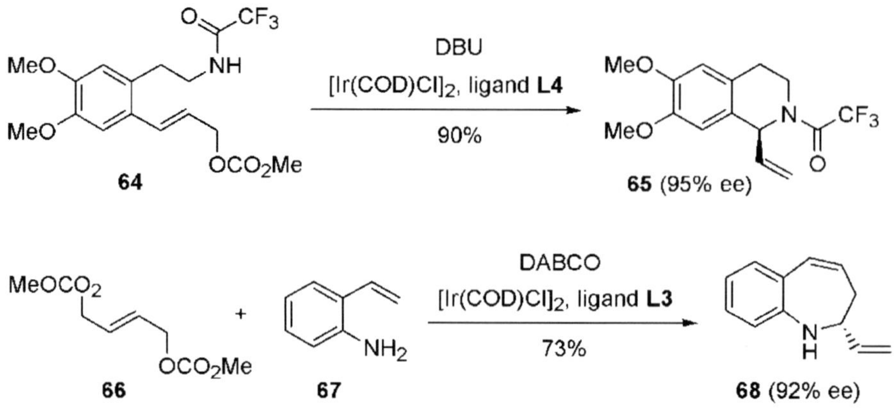

Scheme 20.

4. Reaction with Oxygen Nucleophiles

4.1. Regio- and Enantioselective Reaction

The enantio- and regioselective iridium-catalyzed allylic substitution with oxygen nucleophiles was studied by Hartwig's group [107-110]. The early work was the etherification with phenolates as nucleophiles (Scheme 21) [107]. In the presence of phosphorus amidite ligand *ent-L3*, the reaction of *17* with PhOLi afforded the chiral branched allylic ether (*R*)-*69* with 96% ee. Next, he reported the direct allylic etherification with aliphatic alcohols in the presence of an alkali metal base [109]. For an example, the reaction of *14* with benzyl alcohol gave the branched allylation product *71* with 93% ee. For high conversions, the use of highly powdered K_3PO_4 was important. Additionally, a catalytic amount of an alkyne additive was used to suppress olefin isomerization that forms vinyl ether side products. Recently, kinetic asymmetric allylic substitution of racemic allylic benzoates was also achieved [110].

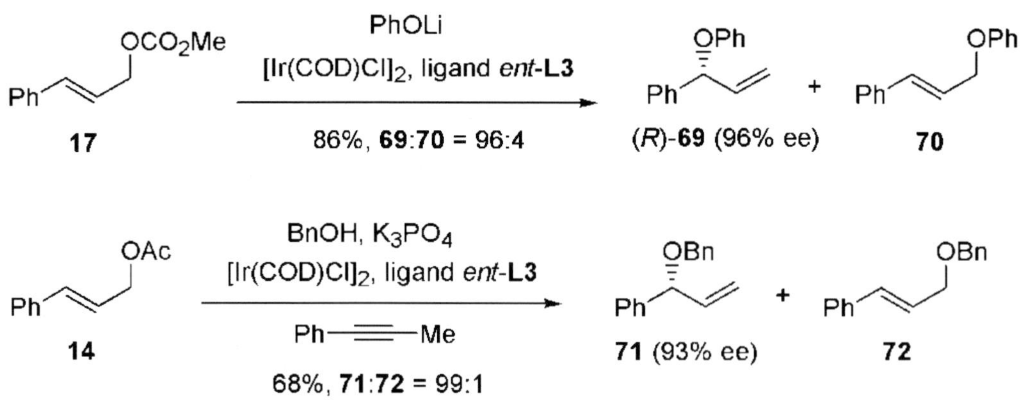

Scheme 21.

Carreira's group reported the iridium-catalyzed kinetic resolution of secondary allylic carbonates using a new class of chiral [2.2.2]-bicyclooctadiene ligands [111]. Treatment of racemic carbonate *73* with phenol and iridium complex, prepared from chiral diene ligand *L11*, led to the formation of the phenyl ether (*S*)-*69* (Scheme 22). The recovered starting carbonate *73* was isolated with 93% ee. He also reported the enantio- and regioselective allylic etherification using potassium silanolates as oxygen nucleophiles [112].

Uozumi's group reported the new P-chiral phosphorodiamidite ligands *L12* and *L13* for iridium-catalyzed allylic etherification (Scheme 23) [113]. In the study on asymmetric allylic etherification of carbonate *17* with phenol, where both (*R*)-*69* and (*S*)-*69* were obtained by changing the *N*- and *P*-substituents of the ligands.

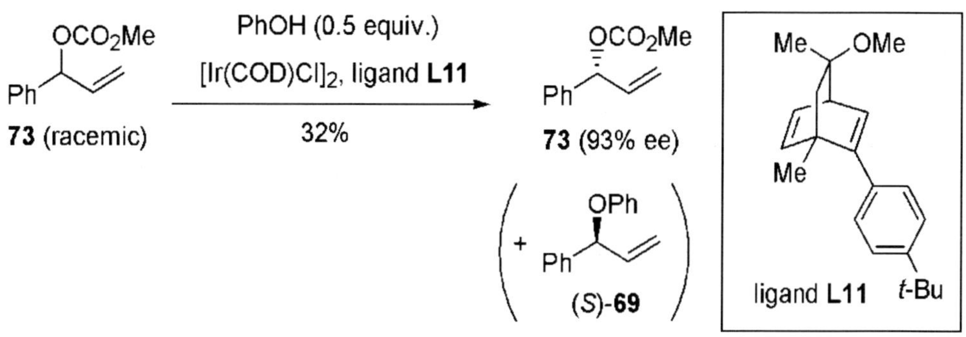

Scheme 22.

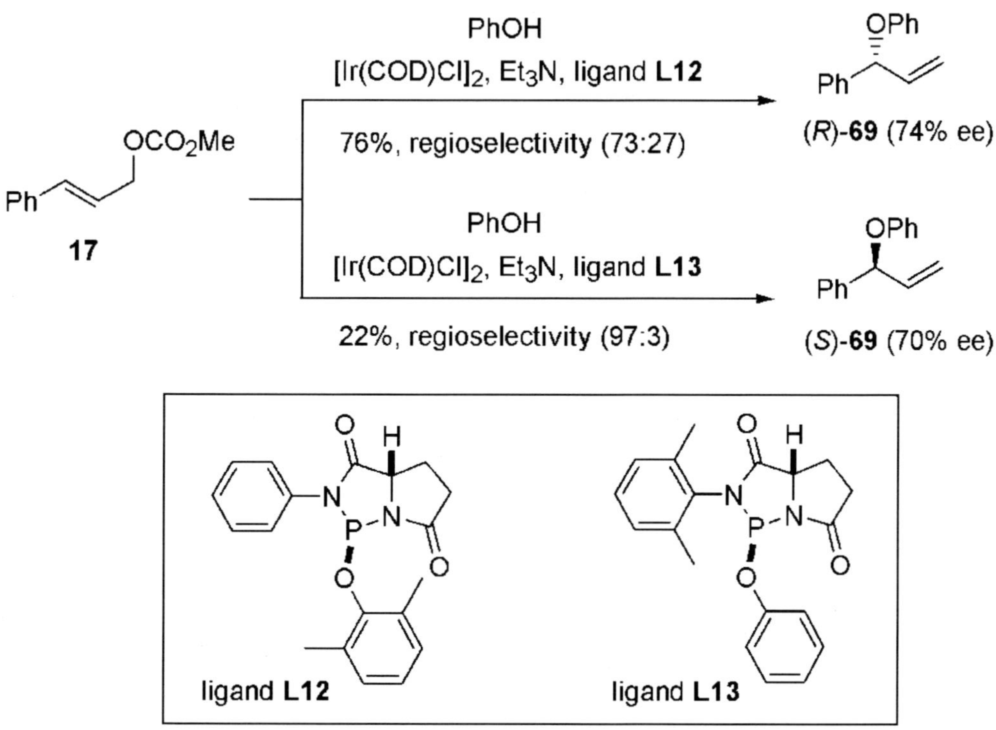

Scheme 23.

Recently, Helmchen's group reported the iridium-catalyzed allylic hydroxylation to give the branched allylic alcohols directly [114]. In this reaction, carbonates such as $KHCO_3$ were used as pronucleophiles.

4.2. Reaction with other Oxygen Nucleophiles

Takemoto and Miyabe found that oximes and hydroxylamines with an *N*-electron-withdrawing substituent, known as hydroxamic acids, act as reactive oxygen nucleophiles [95,115-118]. The iridium-complex of pybox ligand *L8* catalyzed the reaction of phosphate *51* with oxime *74* to form the branched product *75* with good enantioselectivity (Scheme 24) [95,118]. The asymmetric allylic substitution of *21* with hydroxylamine *76* was studied [116]. In the presence of Et_2Zn as base, the regioselective reaction of *78* with hydroxylamine *79* was also investigated for preparing the 1,2-oxazine core structure *81* of FR900482 [117].

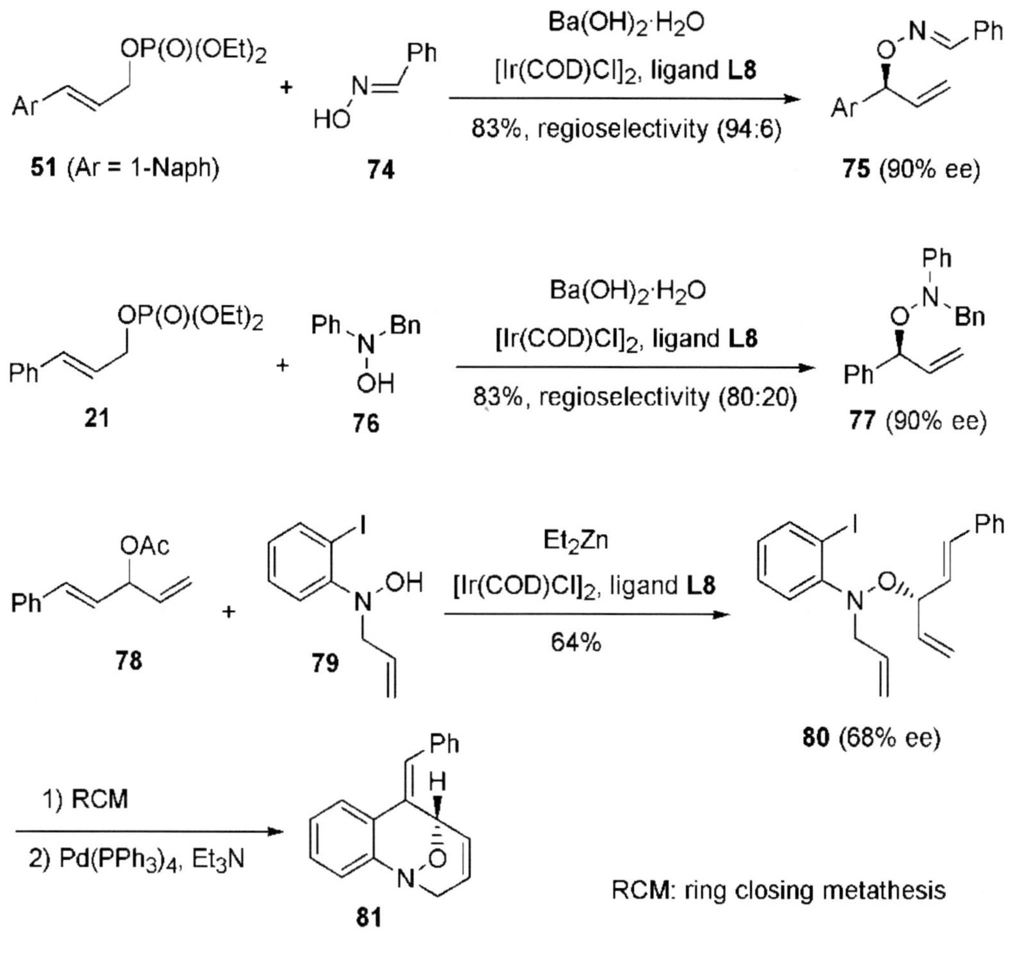

Scheme 24.

4.3. Iridium-Catalyzed Cyclization

Helmchen's group reported the enantioselective intramolecular etherification of carbonate *82* by using the iridium complex with ligand *L4* (Scheme 25) [83]. The good enantioselectivity was obtained upon activation with TBD. The chromane derivative *83* was obtained in 60% yield with 95% ee.

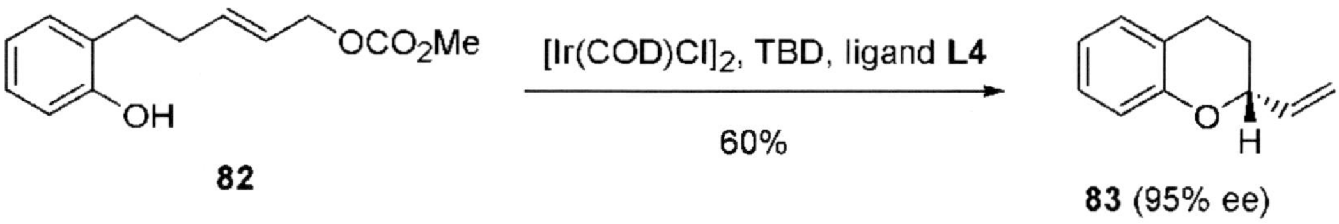

Scheme 25.

5. Reaction with Sulfur Nucleophiles

The construction of a carbon-sulfur bond in iridium-catalyzed allylic substation reactions was explored. With a catalyst *ent-C1*, the allylic substitution with sodium sulfinate was studied by Hartwig's group (Scheme 26) [119]. The reaction of *17* with *84* proceeded in the favor of branched product *85* with 94% ee. The carbon-sulfur bond formation *via* allylic substation with aliphatic thiols was achieved by Zhao's group [120]. In the presence of ligand *L3*, the iridium-catalyzed asymmetric alkylation of sodium cyclohexanethiolate *86* gave the product *87* with 96% ee.

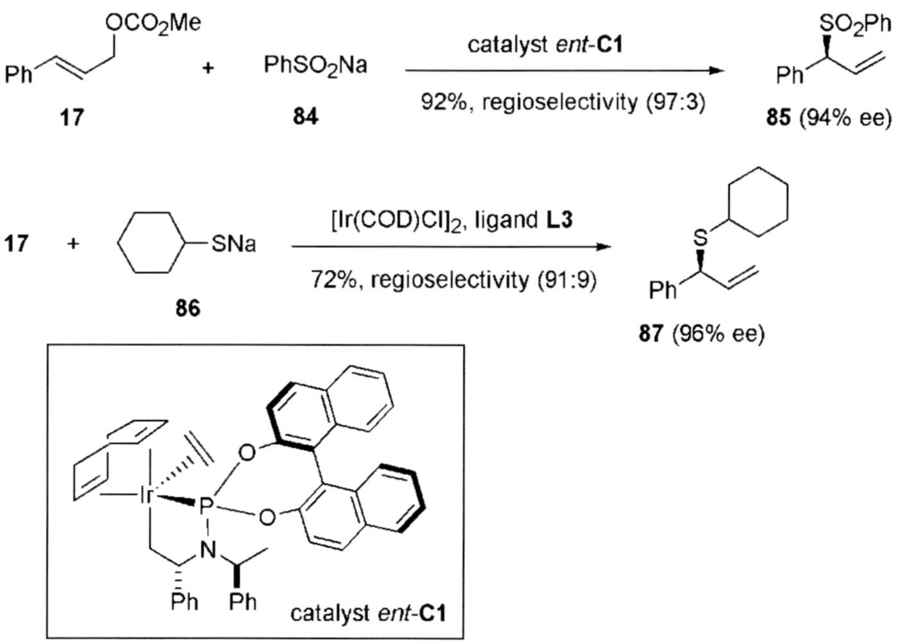

Scheme 26.

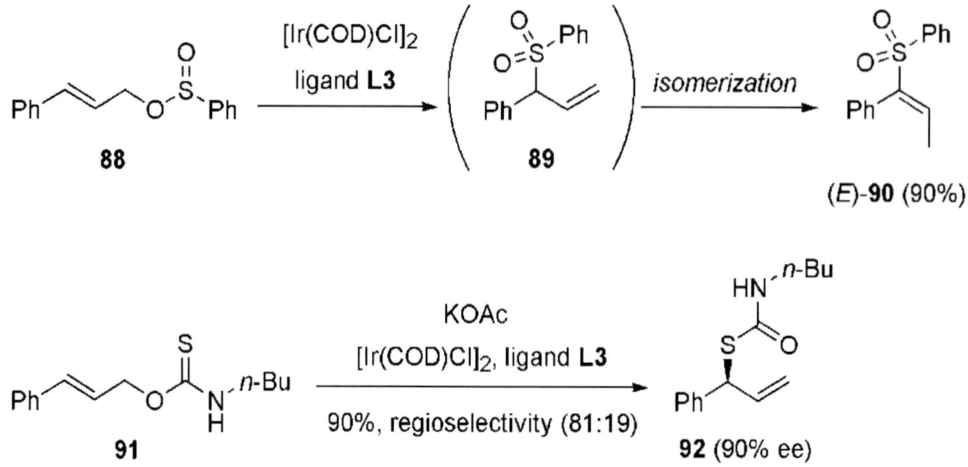

Scheme 27.

You's group also investigated the carbon-sulfur bond formation reactions. The tandem reaction involving iridium-catalyzed allylic substitution of allyl sulfinate *88* and isomerization was reported (Scheme 27) [121]. This reaction allowed ready access to trisubstituted vinyl sulfone (*E*)-*90* exclusively as an *E* isomer. In this transformation, the formation of sulfone (*E*)-*90* is through the isomerization of the allyl sulfone *89*. You's group developed the iridium-catalyzed allylic substitution that afforded chiral *S*-allyl carbamothioates [122]. In the presence of KOAc as base, the allylic substitution of *O*-allyl carbamothioate *31* proceeded smoothly to afford *S*-allyl carbamothioate *92* in 90% ee. In this study, the crossover experiment excludes the rearrangement pathway and supports the π-allyl intermediate pathway.

References

[1] Trost, B.M., Vranken, D.L. (1996) Asymmetric transition-metal-catalyzed allylic alkylations. *Chem. Rev.*, 96, 395-422.

[2] Johannsen, M., Jørgensen, K.A. (1998) Allylic amination. *Chem. Rev.*, 98, 1689-1708.

[3] Helmchen, G. (1999) Enantioselective palladium-catalyzed allylic substitutions with asymmetric chiral ligands. *J. Organomet. Chem.*, 576, 203-214.

[4] Trost, B.M., Crawley, M.L. (2003) Asymmetric transition-metal-catalyzed allylic alkylations: Applications in total synthesis. *Chem. Rev.*, 103, 2921-2944.

[5] Trost. B.M. (2004) Asymmetric allylic alkylation, an enabling methodology. *J. Org. Chem.*, 69, 5813-5837.

[6] Takemoto, Y., Miyabe, H. (2006) C-N bond formation through amination. In R.H. Crabtree, & D.M.P. Mingos (Eds.), *Comprehensive Organometallic Chemistry*, 3rd Edition (pp 695-724). Oxford, Elsevier.

[7] Lu, Z., Ma, S. (2008) Metal-catalyzed enantioselective allylation in asymmetric synthesis. *Angew. Chem. Int. Ed.*, 47, 258-297.

[8] Yoshioka, E., Kohtani, S., Miyabe, H. (2009) Synthesis of heterocycles using allylic substitution with heteroatom nucleophiles. *Trends in Heterocyclic Chemistry*, 14, 1-16.

[9] Takemoto, Y., Miyabe, H. (2010) Asymmetric carbon-heteroatom bond forming reactions. In I. Ojima (Ed.), *Catalytic Asymmetric Synthesis*, 3rd Edition (pp 227-267). UK, John Wiley & Sons.
[10] Hayashi, T., Kishi, K., Yamamoto, A., Ito, Y. (1990) Asymmetric allylic amination catalyzed by chiral ferrocenylphosphine-palladium complexes. *Tetrahedron Lett.*, 31, 1743-1746.
[11] Hayashi, T., Ohno, A., Lu, S.J., Matsumoto, Y., Fukuyo, E., Yanagi, K. (1994) Optically active ruthenocenylbis(phosphines): New efficient chiral phosphine ligands for catalytic asymmetric reactions. *J. Am. Chem. Soc.*, 116, 4221-4226.
[12] Trost, B. M., Krische, M. J., Radinov, R., Zanoni, G. (1996) On asymmetric induction in allylic alkylation via enantiotopic facial discrimination. *J. Am. Chem. Soc.*, 118, 6297-6298.
[13] Trost, B.M., Toste, F.D. (1999) Regio- and enantioselective allylic alkylation of an unsymmetrical substrate: A working model. *J. Am. Chem. Soc.*, 121, 4545-4554.
[14] You, S.L., Zhu, X.Z., Luo, Y.M.. Hou, X.L.. Dai, L.X. (2001) Highly regio- and enantioselective Pd-catalyzed allylic alkylation and amination of monosubstituted allylic acetates with novel ferrocene P,N-ligands. *J. Am. Chem. Soc.*, 123, 7471-7472.
[15] Lüssem, B.J.. Gais, H.J. (2003) Palladium-catalyzed deracemization of allylic carbonates in water with formation of allylic alcohols: Hydrogen carbonate ion as nucleophile in the palladium-catalyzed allylic substitution and kinetic resolution. *J. Am. Chem. Soc.*, 125, 6066-6067.
[16] Hilgraf, R., Pfaltz, A. (2005) Chiral bis(*N*-sulfonylamino)phosphine- and TADDOL-phosphite-oxazoline ligands: Synthesis and application in asymmetric catalysis. *Adv. Synth. Catal.*, 347, 61-77.
[17] Liu, W., Chen, D., Zhu, X.Z., Wan, X.L., Hou, X.L. (2009) Highly diastereo- and enantioselective Pd-catalyzed cyclopropanation of acyclic amides with substituted allyl carbonates. *J. Am. Chem. Soc.*, 131, 8734-8735.
[18] Trost, B.M., Hachiya, I. (1998) Asymmetric molybdenum-catalyzed alkylations. *J. Am. Chem. Soc.*, 120, 1104-1105.
[19] Trost, B.M., Dogra, K., Hachiya, I., Emura, T., Hughes, D.L., Krska, S., Reamer, R.A., Palucki, M., Yasuda, N., Reider, P.J. (2002) Designed ligands as probes for the catalytic binding mode in Mo-catalyzed asymmetric allylic alkylation. *Angew. Chem. Int. Ed.*, 41, 1929-1932.
[20] Glorius, F., Pfaltz, A. (1999) Enantioselective molybdenum-catalyzed allylic alkylation using chiral bisoxazoline ligands. *Org. Lett.*, 1, 141-144.
[21] Belda, O., Moberg, C. (2004) Molybdenum-catalyzed asymmetric allylic alkylations. *Acc. Chem. Res.*, 37, 159-167.
[22] Trost, B.M., Dogra, K., Franzini, M. (2004) 5*H*-Oxazol-4-ones as building blocks for asymmetric synthesis of □-hydroxycarboxylic acid derivatives. *J. Am. Chem. Soc.*, 126, 1944-1945.
[23] Leahy, D.K., Evans, P.A. (2005) Rhodium(I)-catalyzed allylic substitution reactions and their applications to target directed synthesis. In P. A. Evans (Ed.), *Modern rhodium-catalyzed organic reactions* (pp 191-214). Weinheim, Wiley-VCH.
[24] Evans, P.A., Robinson, J.E., Nelson, J.P. (1999) Enantiospecific synthesis of allyl amines via the regioselective rhodium-catalyzed allylic amination reaction. *J. Am. Chem. Soc.*, 121, 6761-6762.

[25] Evans, P.A., Leahy, D.K. (2000) Regioselective and enantiospecific rhodium-catalyzed intermolecular allylic etherification with ortho-substituted phenols. *J. Am. Chem. Soc.*, 122, 5012-513.

[26] Evans, P.A., Leahy, D.K. (2002) Regio- and enantiospecific rhodium-catalyzed allylic etherification reactions using copper(I) alkoxides: Influence of the copper halide salt on selectivity. *J. Am. Chem. Soc.*, 124, 7882-7883.

[27] Hayashi, T., Okada, A., Suzuka, T., Kawatsura, M. (2003) High enantioselectivity in rhodium-catalyzed allylic alkylation of 1-substituted 2-propenyl acetates. *Org. Lett.*, 5, 1713-1715.

[28] Evans, P.A., Leahy, D.K., Andrews, W. J., Uraguchi, D. (2004) Stereodivergent construction of cyclic ethers by a regioselective and enantiospecific rhodium-catalyzed allylic etherification: Total synthesis of gaur acid. *Angew. Chem. Int. Ed.*, 43, 4788-4791.

[29] Kazmaier, U., Stolz, D. (2006) Regio- and stereoselective rhodium-catalyzed allylic alkylations of chelated enolates. *Angew. Chem. Int. Ed.*, 45, 3072-3075.

[30] Evans, P.A., Clizbe, E.A. (2009) Unlocking ylide reactivity in the metal-catalyzed allylic substitution reaction: Stereospecific construction of primary allylic amines with aza-ylides. *J. Am. Chem. Soc.*, 131, 8722-8723.

[31] Vrieze, D.C., Hoge, G.S., Hoerter, P.Z., Haitsma, J.T.V., Samas, B.M. (2009) A highly enantioselective allylic amination reaction using a commercially available chiral rhodium catalyst: Resolution of racemic allylic carbonates. *Org. Lett.*, 11, 3140-3142.

[32] Trost, B.M., Fraisse, P.L., Ball, Z.T. (2002) A stereospecific ruthenium-catalyzed allylic alkylation. *Angew. Chem. Int. Ed.*, 41, 1059-1061.

[33] Mbaye, M.D., Renaud, J.L., Demerseman, B. Bruneau, C. (2004) First enantioselective allylic etherification with phenols catalyzed by chiral ruthenium bisoxazoline complexes. *Chem. Commun.*, 1870-1871.

[34] Onitsuka, K., Okuda, H., Sasai, H. (2008) Regio- and enantioselective *O*-allylation of phenol and alcohol catalyzed by a planar-chiral cyclopentadienyl ruthenium complex. *Angew. Chem. Int. Ed.*, 47, 1454-1457.

[35] Austeri, M., Linder, D., Lacour, J. (2008) Enantioselective and regioselective ruthenium-catalyzed decarboxylative etherification of allyl aryl carbonates. *Chem. Eur. J.*, 14, 5737-5741.

[36] Lloyd-Jones, G.C., Pfaltz, A. (1995) Chiral phosphanodihydrooxazoles in asymmetric catalysis: Tungsten-catalyzed allylic substitution. *Angew. Chem. Int. Ed.*, 34, 462-464.

[37] Enders, D., Finkam, M. (1993) Iron mediated synthesis of 4-amino-enoates of high enantiomeric purity. *Synlett*, 401-402.

[38] Plietker, B. (2006) A highly regioselective salt-free iron-catalyzed allylic alkylation. *Angew. Chem. Int. Ed.*, 45, 1469-1473.

[39] Bricout, H., Carpentier, J.F., Mortreux, A. (1995) Nickel-catalyzed substitution reactions of allylic compounds with soft nucleophiles: an efficient alternative to palladium catalysis. *J. Chem. Soc., Chem. Commun.*, 1863-1864.

[40] Bricout, H., Carpentier, J.F., Mortreux, A. (1998) Synthetic and kinetic aspects of nickel-catalyzed amination of allylic alcohol derivatives. *Tetrahedron*, 54, 1073-1084.

[41] Berkowitz, D.B., Bose, M., Choi, S. (2002) In situ enzymatic screening (ISES): A tool for catalyst discovery and reaction development. *Angew. Chem. Int. Ed.*, 41, 1603-1607.

[42] Berkowitz, D.B., Maiti, G. (2004) Following an ISES lead: The first examples of asymmetric Ni(0)-mediated allylic amination. *Org. Lett.*, 6, 2661-2664.

[43] Takeuchi, R. (2002) Iridium complex-catalyzed highly selective organic synthesis. *Synlett*, 1954-1965.

[44] Miyabe, H., Takemoto, Y. (2005) Regio- and stereocontrolled palladium- or iridium-catalyzed allylation. *Synlett*, 1641-1655.

[45] Takeuchi, R., Kezuka, S. (2006) Iridium-catalyzed formation of carbon-carbon and carbon-heteroatom bonds. *Synthesis*, 3349-3366.

[46] Helmchen, G., Dahnz, A., Dübon, P., Schelwies, M., Weihofen, R. (2007) Iridium-catalysed asymmetric allylic substitutions. *Chem. Commun.*, 675-691.

[47] Helmchen, G. (2008) Iridium-catalyzed asymmetric allylic substitutions. In L.A. Oro, & C. Claver (Eds.), *Iridium complexes in organic synthesis* (pp 211-250). Weinheim, Wiley-VCH.

[48] Takeuchi, R., Kashio, M. (1997) Highly selective allylic alkylation with a carbon nucleophile at the more substituted allylic terminus catalyzed by an iridium complex: An efficient method for constructing quaternary carbon centers. *Angew. Chem. Int. Ed.*, 36, 263-265.

[49] Takeuchi, R., Kashio, M. (1998) Iridium complex-catalyzed allylic alkylation of allylic esters and allylic alcohols: Unique regio- and stereoselectivity. *J. Am. Chem. Soc.*, 120, 8647-8655.

[50] Janssen, J.P., Helmchen, G. (1997) First enantioselective alkylations of monosubstituted allylic acetates catalyzed by chiral iridium complexes. *Tetrahedron Lett.*, 38, 8025-8026.

[51] Bartels, B., Helmchen, G. (1999) Ir-catalyzed allylic substitution: Mechanistic aspects and asymmetric synthesis with phosphorus amidites as ligands. *Chem. Commun.*, 741-742.

[52] Bartels, B., Carcía-Yebra, C., Rominger, F., Helmchen, G. (2002) Iridium-catalysed allylic substitution: Stereochemical aspects and isolation of Ir^{III} complexes related to the catalytic cycle. *Eur. J. Inorg. Chem.*, 2569-2586.

[53] Bartels, B., Carcía-Yebra, C., Helmchen, G. (2003) Asymmetric Ir^{I}-catalyzed allylic alkylation of monosubstituted allylic acetates with phosphorus amidites as ligands. *Eur. J. Org. Chem.*, 1097-1103.

[54] Lipowsky, G., Miller, N., Helmchen, G. (2004) Regio- and enantioselective iridium-catalyzed allylic alkylation with in situ activated P,C-chelate complexes. *Angew. Chem. Int. Ed.*, 43, 4595-4597.

[55] Carcía-Yebra, C., Janssen, J. P., Rominger, F., Helmchen, G. (2004) Asymmetric iridium(I)-catalyzed allylic alkylation of monosubstituted allylic substrates with phosphinooxazolines as ligands. Isolation, characterization, and reactivity of chiral (allyl)iridium(III) complexes. *Organometallics*, 23, 5459-5470.

[56] Streiff, S., Welter, C., Schelwies, M., Lipowsky, G., Miller, N., Helmchen, G. (2005) Carbocycles via enantioselective inter- and intramolecular iridium catalysed allylic alkylations. *Chem. Commun.*, 2957-2957.

[57] Fuji, K., Kinoshita, N., Tanaka, K., Kawabata, T. (1999) Enantioselective allylic substitution catalyzed by an iridium complex: Remarkable effects of the counter cation. *Chem. Commun.*, 2289-2290.

[58] Kinoshita, N., Marx, K. H., Tanaka, K., Tsubaki, K., Kawabata, T., Yoshikai, N., Nakamura, E., Fuji, K. (2004) Enantioselective allylic substitution of cinnamyl esters catalyzed by iridium-chiral aryl phosphite complex: Conspicuous change in the mechanistic spectrum by a countercation and solvent. *J. Org. Chem.*, 69, 7960-7964.

[59] Alexakis, A., Polet, D. (2004) Very efficient phosphoramidite ligand for asymmetric iridium-catalyzed allylic alkylation. *Org. Lett.*, 6, 3529-3532.

[60] Polet, D., Alexakis, A. (2005) Kinetic study of various phosphoramidite ligands in the iridium-catalyzed allylic substitution. *Org. Lett.*, 7, 1621-1624.

[61] Polet, D., Alexakis, A., Tissot-Croset, K., Corminboeuf, C., Ditrich, K. (2006) Phosphoramidite ligands in iridium-catalyzed allylic substitution. *Chem. Eur. J.*, 12, 3596-3609.

[62] Liu, W.B., Zheng, S.C., He, H., Zhao, X.M., Daia, L.X., You, S.L. (2009) Iridium-catalyzed regio- and enantioselective allylic alkylation of fluorobis(phenylsulfonyl) methane. *Chem. Commun.*, 6604-6606.

[63] Kanayama, T., Yoshida, K., Miyabe, H., Takemoto, Y. (2003) Enantio- and diastereoselective Ir-catalyzed allylic substitutions for asymmetric synthesis of amino acid derivatives. *Angew. Chem. Int. Ed.*, 42, 2054-2056.

[64] Kanayama, T., Yoshida, K., Miyabe, H., Kimachi, T., Takemoto, Y. (2003) Synthesis of □–substituted □-amino acids with use of iridium-catalyzed asymmetric allylic substitution. *J. Org. Chem.*, 68, 6197-6201.

[65] Bondzic, B.P., Farwick, A., Liebich, J., Eilbracht, P. (2008) Application of iridium catalyzed allylic substitution reactions in the synthesis of branched tryptamines and homologues via tandem hydroformylation–Fischer indole synthesis. *Org. Biomol. Chem.*, 6, 3723-3731.

[66] Wu, Q.F., He, H., Liu, W.B., You, S.L. (2010) Enantioselective construction of spiroindolenines by Ir-catalyzed allylic alkylation reactions. *J. Am. Chem. Soc.*, 132, 11418-11419.

[67] Takeuchi, R., Shiga, N. (1999) Complete retention of Z geometry in allylic substitution catalyzed by an iridium complex. *Org. Lett.*, 1, 265-267.

[68] Takeuchi, R., Ue, N., Tanabe, K., Yamashita, K., Shiga, N. (2001) Iridium complex-catalyzed allylic amination of allylic esters. *J. Am. Chem. Soc.*, 123, 9525-9534.

[69] Ohmura, T., Hartwig, J.F. (2002) Regio- and enantioselective allylic amination of achiral allylic esters catalyzed by an iridium-phosphoramidite complex. *J. Am. Chem. Soc.*, 124, 15164-15165.

[70] Kiener, C.A., Shu, C., Incarvito, C., Hartwig, J.F. (2003) Identification of an activated catalyst in the iridium-catalyzed allylic amination and etherification. Increased rates, scope, and selectivity. *J. Am. Chem. Soc.*, 125, 14272-14273.

[71] Shu, C., Leitner, A., Hartwig, J.F. (2004) Iridium-catalyzed intermolecular allylic etherification with aliphatic alkoxides: Asymmetric synthesis of dihydropyrans and dihydrofurans. *Angew. Chem. Int. Ed.*, 43, 4797-4800.

[72] Leitner, A., Shekhar, S., Pouy, M.J., Hartwig, J.F. (2005) A simple iridium catalyst with a single resolved stereocenter for enantioselective allylic amination. Catalyst selection from mechanistic analysis. *J. Am. Chem. Soc.*, 127, 15506-15514.

[73] Leitner, A., Shu, C., Hartwig, J.F. (2005) Effects of catalyst activation and ligand steric properties on the enantioselective allylation of amines and phenoxides. *Org. Lett.*, 7, 1093-1096.

[74] Shekhar, S., Trantow, B., Leitner, A., Hartwig, J.F. (2006) Sequential catalytic isomerization and allylic substitution. Conversion of racemic branched allylic carbonates to enantioenriched allylic substitution products. *J. Am. Chem. Soc.*, 128, 11770-11771.

[75] Pouy, M.J., Leitner, A., Weix, D.J., Ueno, S., Hartwig, J.F. (2007) Enantioselective iridium-catalyzed allylic amination of ammonia and convenient ammonia surrogates. *Org. Lett.*, 9, 3949-3952.

[76] Yamashita, Y., Gopalarathnam, A., Hartwig, J.F. (2007) Iridium-catalyzed, asymmetric amination of allylic alcohols activated by Lewis acids. *J. Am. Chem. Soc.*, 129, 7508-7509.

[77] Marković, D., Hartwig, J.F. (2007) Resting state and kinetic studies on the asymmetric allylic substitutions catalyzed by iridium-phosphoramidite complexes. *J. Am. Chem. Soc.*, 129, 11680-11681.

[78] Stanley, L.M., Hartwig, J.F. (2009) Regio- and enantioselective *N*-allylations of imidazole, benzimidazole, and purine heterocycles catalyzed by single-component metallacyclic iridium complexes. *J. Am. Chem. Soc.*, 131, 8971-8983.

[79] Pouy, M.J., Stanley, L.M., Hartwig, J.F. (2009) Enantioselective, iridium-catalyzed monoallylation of ammonia. *J. Am. Chem. Soc.*, 131, 11312-11313.

[80] Weix, D.J., Marković, D., Ueda, M., Hartwig, J.F. (2009) Direct, intermolecular, enantioselective, iridium-catalyzed allylation of carbamates to form carbamate-protected, branched allylic amines. *Org. Lett.*, 11, 2944-2947.

[81] Lipowsky, G., Helmchen, G. (2004) Regio- and enantioselective iridium-catalysed allylic aminations and alkylations of dienyl esters. *Chem. Commun.*, 116-117.

[82] Welter, C., Koch, O., Lipowsky, G., Helmchen, G. (2004) First intramolecular enantioselective iridium-catalysed allylic aminations. *Chem. Commun.*, 896-897.

[83] Welter, C., Dahnz, A., Brunner, B., Streiff, S., Duebon, P., Helmchen, G. (2005) Highly enantioselective syntheses of heterocycles via intramolecular Ir-catalyzed allylic amination and etherification. *Org. Lett.*, 7, 1239-1242.

[84] Weihofen, R., Dahnz, A., Tverskoy, O., Helmchen, G. (2005) Highly enantioselective iridium-catalysed allylic aminations with anionic *N*-nucleophiles. *Chem. Commun.*, 3541-3543.

[85] Welter, C., Moreno, R.M., Streiff, S., Helmchen, G. (2005) Enantioselective synthesis of (+)(*R*)- and (−)(*S*)-nicotine based on Ir-catalyzed allylic amination. *Org. Biomol. Chem.*, 3, 3266-3268.

[86] Weihofen, R., Tverskoy, O., Helmchen, G. (2006) Salt-free iridium-catalyzed asymmetric allylic aminations with N,N-diacylamines and ortho-nosylamide as ammonia equivalents. *Angew. Chem. Int. Ed.*, 45, 5546-5549.

[87] Spiess, S., Berthold, C., Weihofen, R., Helmchen, G. (2007) Synthesis of □,□-unsaturated □-lactams via asymmetric iridium-catalyzed allylic substitution. *Org. Biomol. Chem.*, 5, 2357-2360.

[88] Gnamm, C., Franck, G., Miller, N., Stork, T., Brödner, K., Helmchen, G. (2008) Enantioselective iridium-catalyzed allylic aminations of allylic carbonates with functionalized side chains. Asymmetric total synthesis of (*S*)-vigabatrin. *Synthesis*, 3331-3350.

[89] Gnamm, C., Krauter, C.M., Brödner, K., Helmchen, G. (2009) Stereoselective synthesis of 2,6-disubstituted piperidines using the iridium-catalyzed allylic cyclization as

configurational switch: Asymmetric total synthesis of (+)-241D and related piperidine alkaloids. *Chem. Eur. J.*, 15, 2050-2054.

[90] Gnamm, C., Brödner, K., Krauter, C.M., Helmchen, G. (2009) A configurational switch based on iridium-catalyzed allylic cyclization: Application in asymmetric total syntheses of prosopis, dendrobate, and spruce alkaloids. *Chem. Eur. J.*, 15, 10514-10532.

[91] Spiess, S., Raskatov, J.A., Gnamm, C., Brödner, K., Helmchen, G. (2009) Ir-catalyzed asymmetric allylic substitutions with (phosphoramidite)Ir complexes—Resting states, synthesis, and characterization of catalytically active (□-allyl)Ir complexes. *Chem. Eur. J.*, 15, 11087-11090.

[92] Raskatov, J.A., Spiess, S., Gnamm, C., Brödner, K., Rominger, F., Helmchen, G. (2010) Ir-catalysed asymmetric allylic substitutions with cyclometalated (phosphoramidite)Ir complexes—Resting states, catalytically active (□-allyl)Ir complexes and computational exploration. *Chem. Eur. J.*, 16, 6601-6615.

[93] Farwick, A., Helmchen, G. (2010) Enantioselective total synthesis of (-)-□-kainic acid. *Org. Lett.*, 12, 1108-1111.

[94] Tissot-Croset, K., Polet, D., Alexakis, A. (2004) A highly effective phosphoramidite ligand for asymmetric allylic substitution. *Angew. Chem. Int. Ed.*, 43, 2426-2428.

[95] Miyabe, H., Matsumura, A., Moriyama, K., Takemoto, Y. (2004) Utility of the iridium complex of the pybox ligand in regio- and enantioselective allylic substitution. *Org. Lett.*, 6, 4631-4634.

[96] Miyabe, H., Yoshida, K., Reddy, V.K., Takemoto, Y. (2009) Palladium or iridium-catalyzed allylic substitution of guanidines: Convenient and direct modification of guanidines. *J. Org. Chem.*, 74, 305-311.

[97] Miyabe, H., Moriyama, K., Takemoto, Y. (2011) Regio- and enantioselective allylic substitution with hydroxylamines catalyzed by iridium-complex of bis(oxazolinyl)pyridine. *Chem. Pharm. Bull.*, 59, 714-720.

[98] Nemoto, T., Sakamoto, T., Matsumoto, T., Hamada, Y. (2006) Ir-catalyzed asymmetric allylic amination using chiral diaminophosphine oxides. *Tetrahedron Lett.*, 47, 8737-8740.

[99] Defieber, C., Ariger, M.A., Moriel, P., Carreira, E.M. (2007) Iridium-catalyzed synthesis of primary allylic amines from allylic alcohols: Sulfamic acid as ammonia equivalent. *Angew. Chem. Int. Ed.*, 46, 3139-3143.

[100] Singh, O.V., Han, H. (2007) Iridium(I)-catalyzed regio- and enantioselective decarboxylative allylic amidation of substituted allyl benzyl imidodicarbonates. *J. Am. Chem. Soc.*, 129, 774-775.

[101] Singh, O.V., Han, H. (2007) Iridium(I)-catalyzed stereospecific decarboxylative allylic amidation of chiral branched benzyl allyl imidodicarboxylates. *Org. Lett.*, 9, 4801-4804.

[102] Singh, O.V., Han, H. (2007) Iridium(I)-catalyzed regio- and enantioselective allylic amidation. *Tetrahedron Lett.*, 48, 7094-7098.

[103] Miyabe, H., Yoshida, K., Kobayashi, Y., Matsumura, A., Takemoto, Y. (2003) Synthesis of azacycles based on iridium-catalyzed sequential allylic amination. *Synlett*, 1031-1033.

[104] Teichert, J.P., Fañanás-Mastral, M., Feringa, B.M. (2011) Iridium-catalyzed asymmetric intramolecular allylic amidation: Enantioselective synthesis of chiral

tetrahydroisoquinolines and saturated nitrogen heterocycles. *Angew. Chem. Int. Ed.*, 50, 688-691.

[105] He, H., Liu, W.B., Dai, L.X., You, S.L. (2010) Enantioselective synthesis of 2,3-dihydro-1*H*-benzo [b]azepines: Iridium-catalyzed tandem allylic vinylation/amination reaction. *Angew. Chem. Int. Ed.*, 49, 1496-1499.

[106] Xia, J.B., Liu, W.B., Wang, T.M., You, S.L. (2010) Enantioselective synthesis of 3-azabicyclo [4.1.0]heptenes and 3-azabicyclo [3.2.0]heptenes by Ir-catalyzed asymmetric allylic amination of N-tosyl propynylamine and Pt-catalyzed cycloisomerization. *Chem. Eur. J.*, 16, 6442-6446.

[107] López, F., Ohmura, T., Hartwig, J.F. (2003) Regio- and enantioselective iridium-catalyzed intermolecular allylic etherification of achiral allylic carbonates with phenoxides. *J. Am. Chem. Soc.*, 125, 3426-3427.

[108] Shu, C., Hartwig, J.F. (2004) Enantioselective allylation of aromatic amines after in situ generation of an activated cyclometalated iridium catalyst. *Angew. Chem. Int. Ed.*, 43, 4794-4797.

[109] Ueno, S., Hartwig, J.F. (2008) Direct, iridium-catalyzed enantioselective and regioselective allylic etherification with aliphatic alcohols. *Angew. Chem. Int. Ed.*, 47, 1928-1931.

[110] Stanley, L.M., Bai, C., Ueda, M., Hartwig, J.F. (2010) Iridium-catalyzed kinetic asymmetric transformations of racemic allylic benzoates. *J. Am. Chem. Soc.*, 132, 8918-8920.

[111] Fischer, C., Defieber, C., Suzuki, T., Carreira, E.M. (2004) Readily available [2.2.2]-bicyclooctadienes as new chiral ligands for Ir(I): Catalytic, kinetic resolution of allyl carbonates. *J. Am. Chem. Soc.*, 126, 1628-1629.

[112] Lyothier, I., Defieber, C., Carreira, E.M. (2006) Iridium-catalyzed enantioselective synthesis of allylic alcohols: Silanolates as hydroxide equivalents. *Angew. Chem. Int. Ed.*, 45, 6204-6207.

[113] Kimura, M., Uozumi, Y. (2007) Development of new P-chiral phosphorodiamidite ligands having a pyrrolo [1,2-*c*]diazaphosphol-1-one unit and their application to regio- and enantioselective iridium-catalyzed allylic etherification. *J. Org. Chem.*, 72, 707-714.

[114] Gärtner, M., Mader, S., Seehafer, K., Helmchen, G. (2011) Enantio- and regioselective iridium-catalyzed allylic hydroxylation. *J. Am. Chem. Soc.*, 133, 2072-2075.

[115] Miyabe, H., Matsumura, A., Yoshida, K., Yamauchi, M., Takemoto, Y. (2004) Novel synthesis of *O*-allylated oxime ethers based on allylic substitution with oximes. *Synlett*, 2123-2126.

[116] Miyabe, H., Yoshida, K., Yamauchi, M., Takemoto, Y. (2005) Hydroxylamines as oxygen atom nucleophiles in transition-metal-catalyzed allylic substitution. *J. Org. Chem.*, 70, 2148-2153.

[117] Reddy, V.K., Miyabe, H., Yamauchi, M., Takemoto, Y. (2008). Enantioselective synthesis of [1,2]-oxazinone scaffolds and [1,2]-oxazine core structures of FR900482. *Tetrahedron*, 64, 1040-1048.

[118] Miyabe, H., Matsumura, A., Yoshida, K., Takemoto, Y. (2009) Synthesis of chiral oxime ethers based on regio- and enantioselective allylic substitution catalyzed by iridium-pybox complex. *Tetrahedron*, 65, 4464-4470.

[119] Ueda, M., Hartwig, J.F. (2010) Iridium-catalyzed, regio- and enantioselective allylic substitution with aromatic and aliphatic sulfinates. *Org. Lett.*, 12, 92-94.
[120] Gao, N., Zheng, S., Yang, W., Zhao, X. (2011) Carbon-sulfur bond formation via iridium-catalyzed asymmetric allylation of aliphatic thiols. *Org. Lett.*, 13, 1514-1516.
[121] Xu, Q.L., Dai, L.X., You, S.L. (2010) Tandem Ir-catalyzed allylic substitution reaction of allyl sulfinates and isomerization. *Org. Lett.*, 12, 800-803.
[122] Xu, Q.L., Liu, W.B., Dai, L.X., You, S.L. (2010) Iridium-catalyzed enantioselective allylic substitution of O-allyl carbamothioates. *J. Org. Chem.*, 75, 4615-4618.

In: Iridium: Occurrence, Characteristics and Applications ISBN: 978-1- 62081-681-3
Editors: Chieko Fukui and Maemi Ono

Chapter 6

A NOVEL AMPEROMETRIC TRANSDUCER ELECTRODE WITH IRIDIUM-NIOBIU BINARY ALLOYS

Toru Matsumoto[1,*], Naoaki Sata[1] and Yoko Yamabe-Mitarai[2,‡]
[1]Green Innovation Research Laboratories, NEC Corporation, Japan
[2]Exploratory Materials Research Laboratory for Energy and Environment Platinum group metals Group, NIMS, Sengen Japan

Keywords: Amperometric; Transducer; Ir; Nb; Alloy

1. INTRODUCTION

The use of iridium (Ir) as a material for amperometric electrodes, especially in biosensors, has been reported in several studies [1-5]. Research results have also been reported on the use of other materials [6], including binary alloys [7], for this purpose. However, using such materials is associated with inadequate hydrogen peroxide detection, susceptibility to interference species (e.g., ascorbic acid, uric acid, and acetaminophen), a high base current, and increased complexity in fabricating electrodes, among other problems [1-5]. Hydrogen peroxide is produced by oxidases, and the interference species exist in body fluids. Adequate hydrogen peroxide detection and insusceptibility to interference species are needed in amperometric biosensors. Accordingly, Platinum has almost always been used in the materials of amperometric biosensor electrodes; carbon is an exception [8-12]. Certain binary Ir alloys, i.e., Ir-Nb, Ir-Zr, and Ir-Nb-Zr, have been noted as having high temperature materials [13, 14]. However, the prices of rare metals including Ir have been fluctuating widely because of the global economic crisis and/or because they are mined in limited

[*]Corresponding author T. Matsumoto, Green Innovation Research Laboratories, NEC Corporation, 34 Miyukigaoka, Tsukuba, Ibaraki, 305-8501, Japan. Tel: +81-29-850-1150; Fax: +81-29-856-6136. E-mail address: *t-matsumoto@dt.jp.nec.com.*
[‡] Exploratory Materials Research Laboratory for Energy and Environment Platinum group metals Group, NIMS, Sengen

quantities. This has created a need to use alternative materials in place of Ir in various fields, including biosensor electrode fabrication. In our work, we found that an Ir-23 at.%Nb (Ir-23Nb) binary alloy can oxidize hydrogen peroxide better than any other Ir-Nb binary alloy.

2. EXPERIMENTAL

We prepared electrodes from eight materials: Ir, Nb, and six Ir-Nb binary alloys, i.e., Ir-13 at.%Nb (Ir-13Nb), Ir-17 at.%Nb (Ir-17Nb), Ir-23 at.%Nb (Ir-23Nb), Ir-30 at.%Nb (Ir-30Nb), Ir-43 at.%Nb (Ir-43Nb), and Ir-62 at.%Nb (Ir-62Nb). Table 1 shows the composition, heat-treatment condition, and existing phase for these materials. The base electrodes were fabricated with an arc-melting method, after which we used a diamond cutter to reduce them to approximately 3 mm in diameter and 1 mm in height and measured their electrochemical characteristics.

Table 1.

Composition	Heat-Treatment Condition	Phase
Ir	as cast	fcc single-phase
Ir-13Nb	2473 K, 20 mins	fcc+$L1_2$ two-phase
Ir-17Nb	2173 K, 72 hrs	fcc+$L1_2$ two-phase
Ir-23Nb	2473 K, 20 mins + 1773 K, 24 hrs	fcc+$L1_2$ two-phase
Ir-30Nb	as cast	$L1_2$ single-phase
Ir-43Nb	as cast	α_2 single-phase
Ir-62Nb	as cast	σ single-phase
Nb	as cast	bcc single-phase

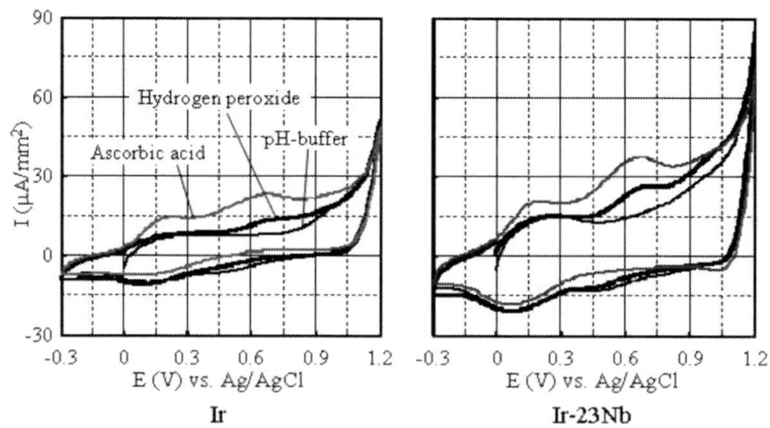

Figure 1. Cyclic voltammograms of Ir and Ir-23Nb binary alloy recorded in pH-buffer (100 mM N-Tris(hydroxymethyl)-methyl-2-aminoethanesulfonic acid, 150 mM sodium Chloride, pH 7). The scan rate was 10 mV/sec., the scan range was between a -0.3 V and 1.2 V versus Ag/AgCl reference electrode. Current increases were observed when 2 mM hydrogen peroxide or 10 mM ascorbic acid was added to the pH-buffer. Ir-23Nb binary alloy was more effective than Ir for oxidizing both hydrogen peroxide and ascorbic acid.

Cyclic voltammetry measurements were used to ascertain the electrochemical behaviors of the pH-buffer, hydrogen peroxide, and ascorbic acid for the eight working electrodes fabricated with the eight materials. The pH-buffer (electrolyte water, pH 7.0) consists of 0.1 M N-Tris(hydroxyl-methyl)-methyl-2-aminoethanesulfonic acid and 0.15 M sodium chloride. Hydrogen peroxide of 2 mM was prepared with the pH-buffer and 30% hydrogen peroxide. Ascorbic acid of 10 mM was prepared with the pH-buffer and an ascorbic acid reagent. Measurements were taken of their cyclic voltammograms to the pH-buffer, the 2 mM hydrogen peroxide, and the 10 mM ascorbic acid using a three-electrode method comprising one of the eight working electrodes, an Ag/AgCl reference electrode, and a Pt counter electrode.

3. Results and Discussion

Figure 1 shows cyclic voltammograms of Ir (left graph) and Ir-23Nb (right graph). With pH-buffer cyclic voltammograms, the Ir oxidation current increased in the potential range from 0 to 0.2 V, reached a plateau between 0.2 and 0.75 V, and soared between 0.75 and 1.2 V. The Ir reduction current was roughly stable between 0 and -0.3 V. In contrast, the Ir-23Nb oxidation current increased more than the Ir oxidation current from 0 to 0.25 V, decreased slightly between 0.25 and 0.5 V, and finally gradually increased more than the Ir oxidation current between 0.5 and 1.2 V. Like the Ir reduction current, the Ir-23Nb reduction current remained stable between 0 and -0.3 V but increased more than the Ir reduction current after that. Due to pH-buffer reactions, increases in oxidation and reduction current occurred in a binary alloy having a 77Ir: 23Nb ratio or in one wherein a phase change from a single-phase fcc to a two-phase fcc+$L1_2$ occurred.

With hydrogen peroxide cyclic voltammograms, the Ir oxidation current slightly increased in the potential range from 0 to 0.2 V, reached an approximate plateau between 0.2 and 0.5 V, increased slightly again between 0.5 and 0.91 V, and soared between 0.91 and 1.2 V. Significant hydrogen peroxide oxidation was observed between 0.45 and 0.95 V. The Ir reduction current remained stable between 0 and -0.3 V, just as it did with the pH-buffer cyclic voltammograms. In contrast, the Ir-23Nb oxidation current gradually increased between 0 and 0.2 V, reached a plateau from approximately 0.2 to over 0.35 V, sharply increased from around 0.45 to 0.7 V, became stable between 0.7 and 0.8 V, and soared more than the Ir oxidation current did between 0.8 and 1.2 V. Significant hydrogen peroxide oxidation was evident between 0.4 and 1.1 V. The Ir-23Nb reduction current increased slightly from 0 to -0.3 V, in contrast with the pH-buffer. In terms of both potential range and capability, this Ir-23Nb hydrogen peroxide oxidation performed significantly better than Ir. Due to hydrogen peroxide and pH-buffer reactions, oxidation and reduction current increases occurred in the binary alloy having a 77Ir: 23Nb ratio or one in which a phase change from a single-phase fcc to two-phase fcc+$L1_2$ occurred.

With ascorbic acid cyclic voltammograms, the Ir oxidation current increased in the potential range from 0 to over 1.2 V, showing two peaks at 0.2 and 0.68 V. The Ir reduction current was largely stable between and 0 and –0.3 V. However, the Ir-23Nb oxidation current also increased from 0 to over 1.2 V, showing two peaks at 0.2 and 0.68 V in the same way as with the Ir oxidation current. Little reduction in either the Ir or Ir-23Nb current occurred. Due to ascorbic acid, pH-buffer, and hydrogen peroxide reactions, oxidation current increases occurred in an binary alloy having a 77Ir: 23Nb ratio or one in which a phase change from a single-phase fcc to a two-phase fcc+$L1_2$ occurred.

We could not determine how a much larger background indicating a higher capacitance will affect the Ir-23Nb electrode's performance, judging from these data. However, the biosensors using this electrode were found to have a short lifetime because enzymes and antibodies by using the biosensors appear to be easily oxidized by the much larger background, indicating a much larger oxidation current.

Figures. 2 (a) and (b) show pH-buffer, 2 mM hydrogen peroxide (HP), and 10 mM ascorbic acid (AA) current dependence on Nb concentration at the potential of 0.7 V for the eight materials used. The currents of the materials themselves were calculated from their cyclic voltammogram levels, and the 0.7 V potential was determined because the HP currents of the materials showed peaks around this voltage level. Figure 2 (a) shows the current dependence of the materials themselves on the Nb concentration, and Figure 2 (b) shows the current dependence of the HP and AA output which was deducted from the 8 element currents of the HP and the AA to the 8 element currents of the pH-buffer on the Nb concentrations. As shown in Figure 2 (a), increasing the Nb concentration to just under 17% Nb (Ir-17Nb) increased the currents of the pH-buffer, the HP, and the AA: The pH-buffer current increased from 8.2 to 48.4 $\mu A/mm^2$, the HP current from 13.5 to 53.5 $\mu A/mm^2$, and the AA current from 23.3 to 67.0 $\mu A/mm^2$. However, increasing the Nb concentration from 17% to 100% decreased the pH-buffer current to approximately 0.001 $\mu A/mm^2$, the HP current to 0.13 $\mu A/mm^2$, and the AA current to 0.12 $\mu A/mm^2$.

As shown in Figure 2 (b), the materials' output currents peaked at 23% Nb (Ir-23Nb): at 9.2 $\mu A/mm^2$ for HP and 20.2 $\mu A/mm^2$ for AA. From this, we calculated the output currents for the following materials with HP and AA: Ir, 5.3 and 15.1 $\mu A/mm^2$; Ir-17Nb, 5.1 and 18.6 $\mu A/mm^2$; Ir-13Nb, 3.7 and 11.7 $\mu A/mm^2$; Ir-30Nb, 2.0 and 5.9 $\mu A/mm^2$; Ir-43, 1.5 and 3.9 $\mu A/mm^2$; Ir-62Nb, 0.6 and 4.0 $\mu A/mm^2$; Nb, 0.13 and 0.12 $\mu A/mm^2$. These results indicate that the Ir-Nb binary alloys having an fcc+$L1_2$ two-phase structure oxidize more effectively than any other Ir-Nb alloys having other phases. The Ir-23Nb alloy has the highest oxidation capability among the fcc+$L1_2$ two-phase alloys. The volume fractions of the $L1_2$ phase in the fcc+$L1_2$ two-phase Ir-Nb alloys are 15, 46, and 92% for the Ir-13Nb, Ir-17Nb, and Ir-23Nb alloys, respectively. The oxidation capability of the Ir-30Nb alloy having 100% of the $L1_2$ phase is lower than the oxidation capability of the fcc+$L1_2$ two-phase alloys. Our results indicate that the volume fraction of the $L1_2$ phase or the metallographic microstructure of the two-phase alloys affects the oxidation capability. Differences in alloy phase structures reportedly affect the oxidation and reduction in hydrogen, oxygen, and HP [15, 16]. However, these reports refer only to Pt binary alloys; no such research of this type has been done on Ir binary alloys. That is to say, the differences in phase structure between these two types of binary alloys could account for the differences in their HP and AA oxidation capacities. Furthermore, whereas the backgrounds (the pH-buffer currents) increase the Ir, the Ir-23Nb, and the Ir-17Nb, in order from most to least, in figure 2(a), the HP output currents increase the Ir-17Nb, the Ir, and the Ir-23Nb, in order from most to least. The AA output currents increase the Ir, the Ir-17Nb, and the Ir-23Nb, in order from most to least in figure 2 (b). The results of these orders indicate that the increase in response is not only due to the increase in the surface area but also due to the increase in the oxidation capability for HP and AA.

a)

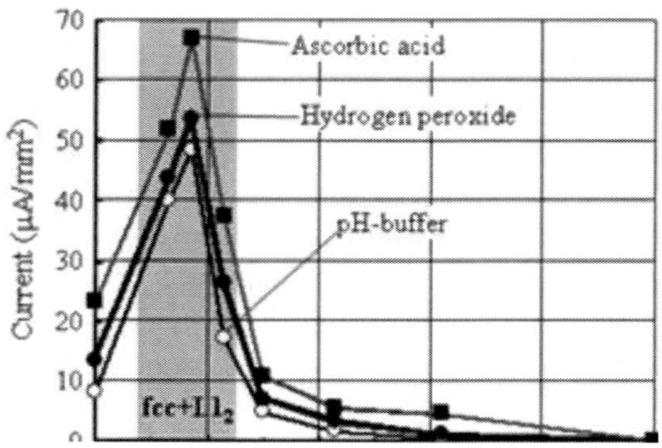

b)

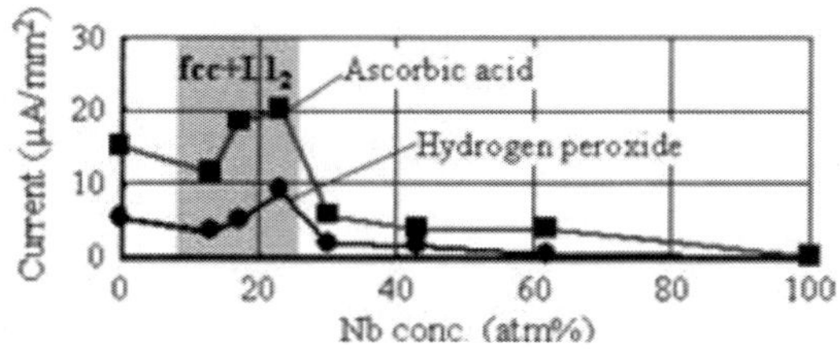

Figure 2. pH-buffer, 2 mM hydrogen peroxide, and 10 mM ascorbic acid current dependence on Nb concentration for eight materials at 0.7 V potential, as obtained from cyclic voltammograms shown in Figure. 1.

(a) is the materials' own currents.

(b) is the ascorbic acid output current and hydrogen peroxide output current, as differentiated from (1) ascorbic acid current and pH-buffer current and (2) hydrogen peroxide current and buffer current.

In previous studies, the constituent phase and the microstructure of both the Ir-13Nb and the Ir-17Nb alloys were investigated using X-ray diffractometry and transmission electron microscopy [17, 18]. The fcc and $L1_2$ two phases clearly appeared in the Ir-13Nb and Ir-17Nb alloys. Although phase identification was not performed for other alloys in our study, the constituent phase could be determined from the Ir-Nb binary phase diagram, and the results are shown in table 1 [19]. The composition range of the fcc+$L1_2$, $L1_2$, α_2, and σ phases is more than 5 at%. Thus, the nominal compositions of 23, 30, 43, 62 at%Nb are located in the fcc+$L1_2$, $L1_2$, α_2, and σ phase region even for a weight loss of less than 2% of the arc-melted samples.These results show that the Ir-23Nb binary alloy can be used instead of Ir as an amperometric biosensor electrode material. Furthermore, it might be used instead of Pt as a transducer material used both of a blood glucose sensor [20] or a urine glucose sensor [9].

Acknowledgments

The authors appreciate the chapter review by Renato Seeber (email: renato.seeber@ unimore.it), Dipartimento di Chimica, Università di Modena e Reggio Emilia, via G. Campi, 183, 41100 Modena, Italy.

CONCLUSION

We found that an Ir-23Nb binary alloy having a two-phase fcc+$L1_2$ more effectively oxidizes hydrogen peroxide and ascorbic acid than Ir having a single-phase fcc, Ir-13Nb having a two-phase fcc+$L1_2$, Ir-17Nb having a two-phase fcc+$L1_2$, Ir-23Nb having a two-phase fcc+$L1_2$, Ir-30Nb having a single-phase $L1_2$, Ir-43Nb having a single-phase α_2, Ir-62Nb having a single-phase σ, or Nb having a single-phase bcc. That is, the Ir-Nb binary alloy having an Nb concentration of around 23% has the most effective oxidation capability of these materials. Our results also showed that the oxidation capability depends not only on the Nb concentration but also on the phase structure of the binary alloy. Thus, we conclude that Ir-23Nb binary alloy electrodes can be applied to amperometric biosensor transducers instead of Ir electrodes.

REFERENCES

[1] Erfan Abu Irhayem, Hanna Elzanowska, Amit S. Jhas, Beata Skrzynecka, and Viola Birss, *Journal of Electroanalytical Chemistry* 2002, 538-539, 153-164.

[2] Faming Tian and Guoyi Zhu, *Sensors and Actuators* B. 2002, 86, 266-270.

[3] Hanna Elzanowska, Erfan Abu-irhayem, Beata Skrzynecka, and Viola I. Birss, *Electroanalysis* 2004, 16, 6 478-490.

[4] Chiak Terashima, Tata N. Rao, Bulusu V. Sarada, Nicolae Spataru, and Akira Fujishima, *Journal of Electroanalytical Chemistry* 2003, 544, 65-74.

[5] James A. Cox and Krzysztof Lewinski, *Talanta* 1993, 40, No. 12, 1911-1915.

[6] Qingling Yang, Plamen Atanasov, and Ebtisam Wilkins, *Sensors and Actuators* B. 1998, 46, 249-256.

[7] K. Yao and Y. F. Cheng, *Journal of Power Sources* 2007, 173, 96-101.

[8] T. Hoshi, T. Noguchi, J. Anzai, *Materials Science and Engineering C.* 2006, 26100-103.

[9] T. Matsumoto, S. Satio, and S. Ikeda, *Journal of Biotechnology* 2006, 122, 267-273.

[10] H. Suzuki, T. Tokuda, T. Miyagishi, H. Yoshida, and N. Honda, *Sensors and Actuators B.* 2004, 97, 90-97.

[11] G. Piechotta, J. Albers, and R. Hintsche, *Biosensors and Bioelectronics* 2005, 21, 802-808.

[12] Y. Zoua, C. Xiang, Li-Xian Suna, and F. Xua, *Biosensors and Bioelectronics* 2008, 23, 1010-1016.

[13] Y. Yamabe-Mitarai and H. Harada, *Journal of Materials Science* 2003, 38, 3955-3963.

[14] Y. Yamabe-Mitarai, H. Harada, Y. Gu, and C. Huang, *Metallurgical and Materials Transactions A* 2005, 36A., 547-557.

[15] V. R. Stamenkovic, B. Fowler, B. S. Mun, G. Wang, P. N. Ross, C. A. Lucas, and N. M. Markovic, *Science* 2007, 315, 493-496.

[16] V. R. Stamenkovic, B. S. Mun, G. M. Arenz, K. J. J. Mayrhofer, C. A. Lucas, G. Wang, P. N. Ross, and N. M. Markovic, *Nature Materials* 2007, 6, 241-247.

[17] Y. Yamabe-Mitarai, Y. Ro, T. Maruko, T. Yokokawa and H. Harada, *Structural Intermetallics* 1997, Edited by M. V. Nathal, R. Caralia, C. T. Liu, and P. L. Martin, The Minerals, Metals and Materials Society, 805-814.
[18] Y. Yamabe-Mitarai and H. Harada, *Philosophical Magazine Letters* 2002, 82, 3, 109-118.
[19] "Phase diagrams for binary alloys", H. Okamoto, *A. S. M.* 2000, 498.
[20] Bao-Yan Wua, Shi-Hua Houb, Feng Yina, Jing Lia, Zi-Xia Zhaoa, Jia-Dong Huanga and Qiang Chena, 2007, 22, 6,838-844.

In: Iridium: Occurrence, Characteristics and Applications
Editors: Chieko Fukui and Maemi Ono
ISBN: 978-1- 62081-681-3

Chapter 7

HIGHLY EFFICIENT ELECTROPHOSPHORESCENT ORGANIC LIGHT-EMITTING DIODES BASED ON IRIDIUM(III) COMPLEXES CONTAINING CARBAZOLE UNITS

Wai-Yeung Wong* and Cheuk-Lam Ho
Hong Kong Baptist University, Waterloo Road, Kowloon Tong,
Hong Kong, P.R. China

ABSTRACT

Since the seminal work on electrophosphorescent organic light-emitting diodes (PHOLEDs) based on fac-[Ir(ppy)3] (Hppy = 2-phenylpyridine), considerable effort has been devoted to the molecular design and synthesis of efficient phosphorescent heavy metal complexes in the fabrication of high-efficiency PHOLEDs, owing to their potential to harness the energies of both the singlet and triplet excitons in the electroluminescence process. Although the emission efficiency and the color of iridium-based PHOLEDs can be readily tuned by structural modifications of the ligand chromophores, the performance of PHOLEDs is largely governed by the charge balance between the electrons and holes from opposite electrodes. It is vital to develop electrophosphors that can show improved charge-balancing features and permits a complete energy transfer between the host and the dopant in the device.

The integration of hole-transporting carbazole module and phosphorescent iridium(III) cyclometalate core appears to be a good choice for the design of more efficient devices. We describe here the synthesis and device performance of a series of interesting small-molecule cyclometalated iridium(III) complexes functionalized with various carbazole units.

Keywords: Carbazole; Electrophosphorescence; Iridium; Organic light-emitting diodes

* Correspondence concerning this article should be addressed to Wai-Yeung Wong, Department of Chemistry and Centre for Advanced Luminescence Materials, Hong Kong Baptist University, Waterloo Road, Kowloon Tong, Hong Kong, P.R. China. E-mail: *rwywong@hkbu.edu.hk*.

ABBREVIATIONS AND SYMBOLS

The following abbreviations and symbols are used throughout the text.

acac	acetylacetonate.
Alq_3	tris(9-hydroxyquinolinato)aluminum.
BCP	2,9-dimethyl-4,7-diphenyl-1,10-phenanthroline.
BPhen	4,7-diphenyl-1,10-phenanthroline.
CBP	4,4'-*N*,*N*'-dicarbazolebiphenyl.
CIE	Commission Internationale de L'Eclairage.
(TD)DFT	(time-dependent) density functional theory.
EL	electroluminescence.
EML	emissive layer.
ETL	electron-transporting layer.
FAB	fast-atom bombardment.
HFlpy	2-(9,9-diethylfluoren-2-yl)pyridine.
fwhm	full width at half maximum.
HI	hole injection.
HT	hole transport.
HOMO	highest occupied molecular orbitals.
HTL	hole-transporting layer.
Hpiq	1-phenylisoquinoline.
Hppy	2-phenylpyridine.
IP	ionization potential.
ISC	intersystem crossing.
L	luminance.
LUMO	lowest unoccupied molecular orbitals.
MLCT	metal to ligand charge transfer.
NPB	4,4'-bis[*N*-(1-naphthyl)-*N*-phenylamino]biphenyl.
NMR	nuclear magnetic resonance.
OLED	organic light-emitting diodes.
PEDOT:PSS	poly(3,4-ethylenedioxythiophene):poly(styrene sulfonate).
PEQ	2-((*E*)-2-phenyl-1-ethenyl)quinoline.
PHOLED	electrophosphorescent organic light-emitting diodes.
PL	photoluminescence.
PVK	poly(*N*-vinylpyridine).
T_{decomp}	onset decomposition temperature.
TGA	thermal gravimetric analysis.
TLC	thin-layer chromatography.
TPBI	2,2',2''-(1,3,5-phenylene)tris(1-phenyl-1*H*-benzimidazole).
UV	ultraviolet.
$V_{turn-on}$	turn-on voltage.
WOLED	white organic light-emitting diodes.
η_{ext}	external quantum efficiency.
η_L	luminance (or current) efficiency.
η_P	power efficiency.

1. INTRODUCTION

Since the seminal work on green organic light-emitting diodes (OLEDs) using *fac*-[Ir(ppy)$_3$] (Hppy = 2-phenylpyridine), [1] considerable effort has been directed towards the development of efficient phosphorescent materials in full-color flat-panel display technology. [2] Among these, cyclometalated iridium(III) complexes represent one of the most studied class of these compounds. [3]

In OLEDs, electrons and holes are injected from opposite electrodes and combine to form spin-singlet or spin-triplet excitons. Radiative decay from the singlet (fluorescence) is fast whereas that from the triplet (phosphorescence) is formally forbidden by the requirement of spin conservation.

Therefore, the introduction of heavy metal atoms is regarded as a novel approach to render the phosphorescence partially allowed, so that both the electrogenerated singlet and triplet excitons can be harvested to give high electroluminescence (EL) efficiency. [4]

While tuning of the emission colors of these heavy metal complexes over the visible spectrum is generally dependent on the choice of cyclometalating ligand or its substituent groups and also on the selection of the auxiliary ligands, [5] the device performance is influenced significantly by the charge balance between electrons and holes from opposite electrodes.

Hence, the charge balance of electrons and holes is becoming an important issue for the new advance of OLED research. Along this line, many researchers are interested in developing some multi-component systems capable of performing separate functional roles in a single molecule. [6]

It is hoped that the next-generation phosphor can exhibit improved charge-balancing features and allow a complete energy transfer between the host and dopant in OLEDs. Many carbazole derivatives such as poly(9-vinylcarbazole) (PVK) and 4,4'-*N,N*'-dicarbazolebiphenyl (CBP) are known to predominantly transport positive charge carriers in organic systems and are high-mobility hole-transporting (HT) materials with a tunable and high triplet energy level. [7,8]

They are also widely used as the host materials for electrophosphors of different emission colors. [8] In order to establish multi-component organometallic systems necessary for more efficient charge transport in the EL process, a combination of HT carbazole moiety and phosphorescent iridium functional core should be a wise choice to afford complexes for efficient electrophosphorescence. The relatively high charge carrier mobility, good thermal stability and inexpensive raw materials also make carbazole-based materials very attractive for practical optoelectronic applications.

Here, we describe our efforts in developing some robust iridium(III) cyclometalated phosphors of green, orange and red emission colors based on various carbazole-derived ligands.

These metal complexes typically possess dual functional roles: light emission and hole transportation, and they show improved charge-transporting properties and morphological stability.

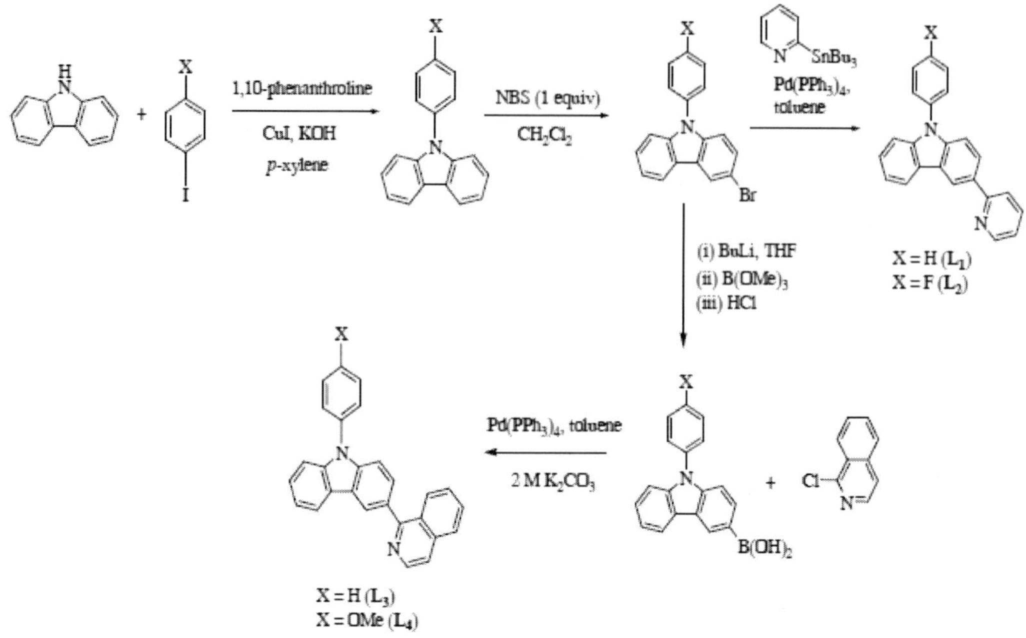

Scheme 1.

2. Multifunctional Iridium(III) Complexes Functionalized with Carbazole Ligands

2.1. Synthesis and Chemical Characterization

The cyclometalating ligand precursors L_1–L_6, which are crucial for the synthesis of all the carbazole-containing iridium(III) cyclometalated complexes 1–10, are prepared following the synthetic routes shown in Schemes 1 and 2. [9] We can synthesize the homoleptic complexes 1-4 and 7–8 in moderate to good yields via a one-step cyclometalation of the appropriate ligand with $[Ir(acac)_3]$ (Hacac = acetylacetone) at high temperature (> 200 °C) in refluxing glycerol (Schemes 3 and 4).

The synthesis of heteroleptic complexes 5, 6, 9 and 10 was accomplished in a two-step sequence, in which the chloride-bridged dimer formed by the reaction of an excess ligand with $IrCl_3{\cdot}nH_2O$ in the initial step can be transformed to the corresponding monomeric complex by replacing the bridging chlorides with acac anion in the presence of Na_2CO_3. Purification of the mixture by preparative thin-layer chromatography (TLC) on silica furnished 1–10 as air-stable powders in high purity.

Each of the iridium(III) phosphors was fully characterized by NMR spectroscopy and FAB mass spectrometry. The relative simplicity of the NMR spectra, which indicates that the number of coupled spins is equal to that of protons on one ligand due to the inherent C_3 symmetry of the triply cyclometalated complexes, confirms the facial geometry around the Ir^{III} center for the homoleptic species. The single-crystal X-ray structures for 1, 3 and 5 were also determined to establish their molecular geometry (vide infra).

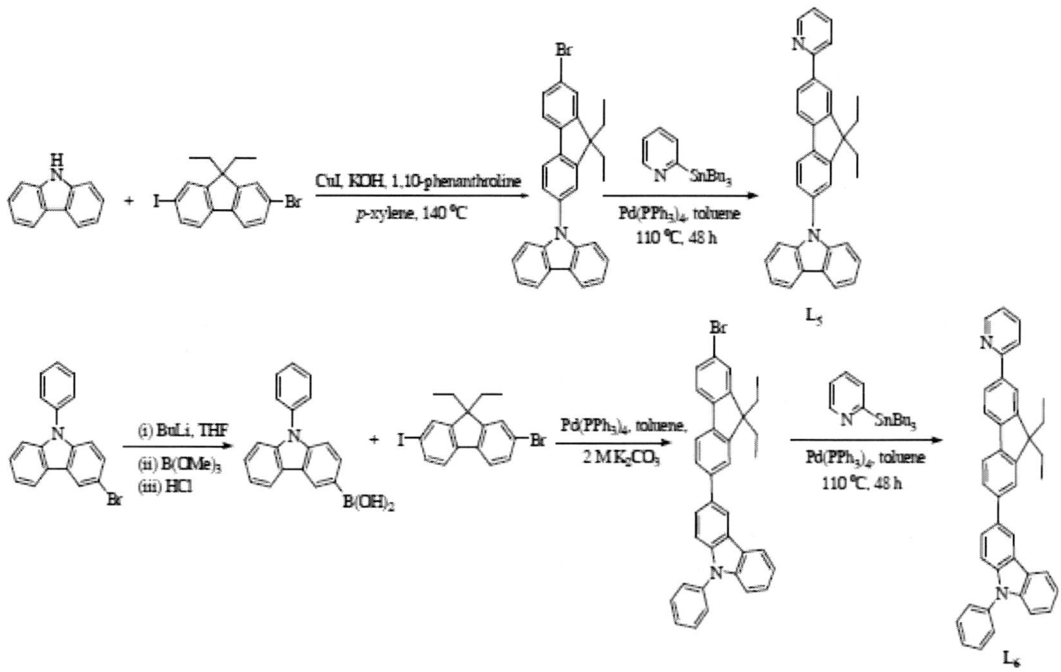

Scheme 2.

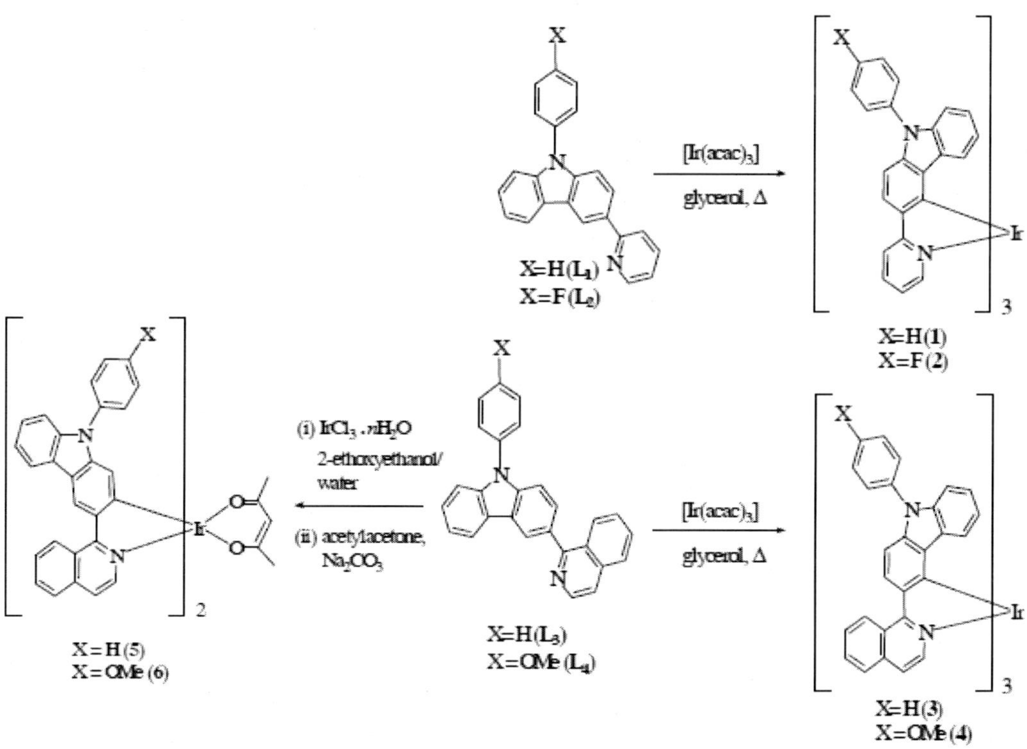

Scheme 3.

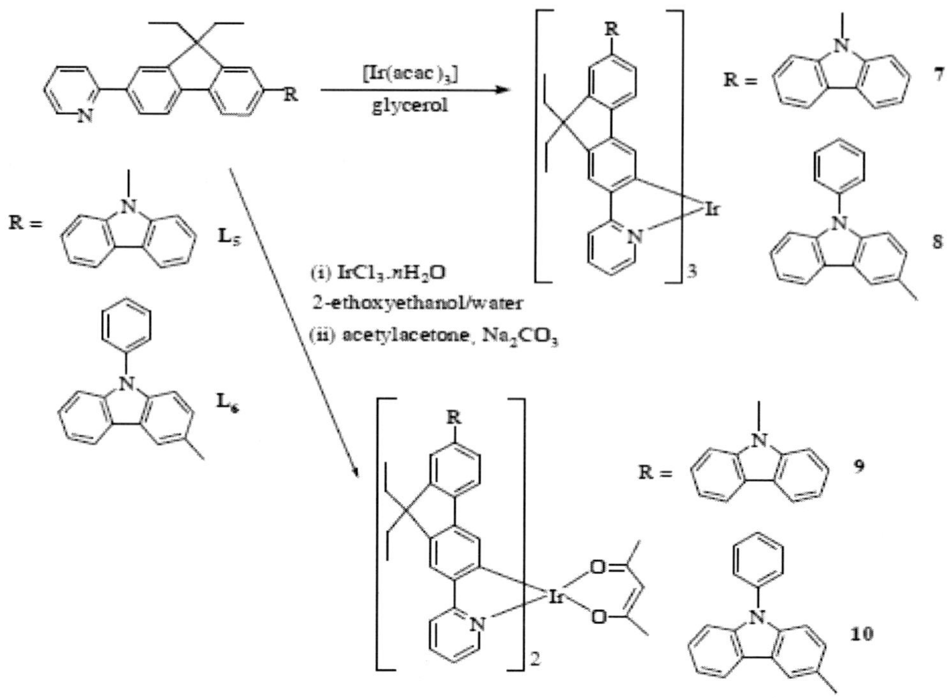
[Ir(acac)$_3$]
glycerol
R =
7
8
L$_5$
L$_6$
(i) IrCl$_3$.nH$_2$O
2-ethoxyethanol/water
(ii) acetylacetone, Na$_2$CO$_3$
R =
9
10

Scheme 4.

a)

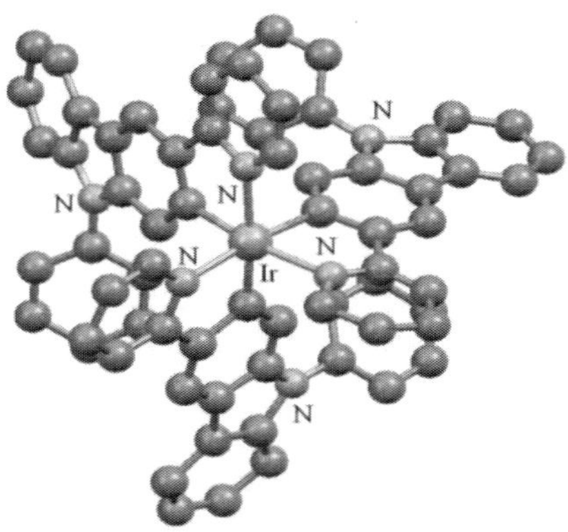
N
N
N
N
N
Ir
N

b)

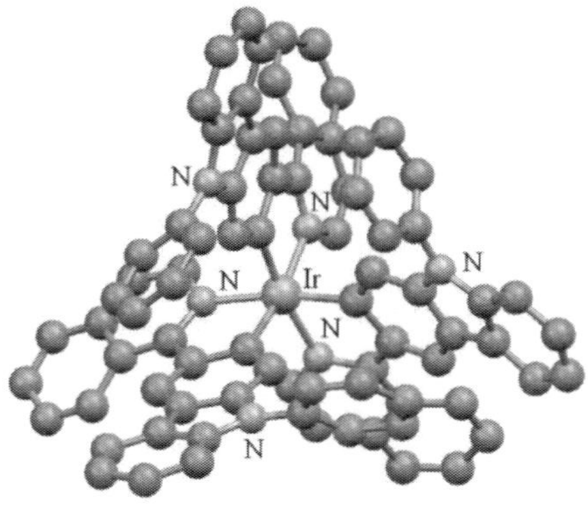
N
N
N
Ir
N
N
N

c)

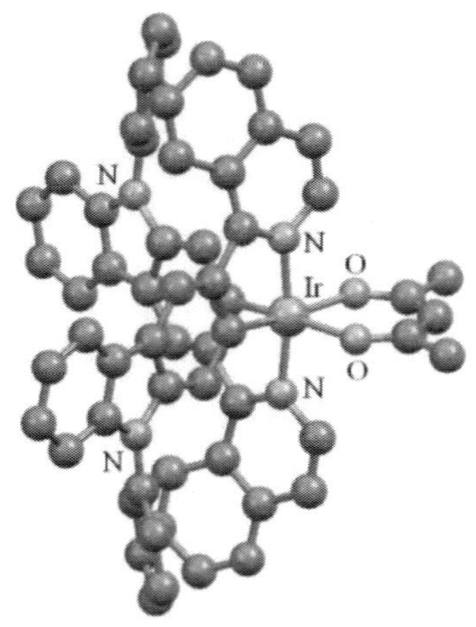

Figure 1. Perspective views of (a) 1, (b) 3 and (c) 5. All carbon atom labels and hydrogen atoms are omitted for clarity.

2.2. Crystal Structures

The X-ray crystal structures of 1, 3 and 5 are shown in Figure 1. The iridium center of each of 1 and 3 adopts a pseudo-octahedral coordination geometry with a facial arrangement sitting on a three-fold axis. Due to the bulky cyclometalating ligands, the Ir–C bonds are found to be shorter than the Ir–N bonds. For 5, the Ir(III) ion shows a distorted octahedral geometry with the *cis*-O,O, *cis*-C,C and *trans*-N,N chelate disposition, and the angle of the *trans*-N,N ligands at the metal center is 172.59(13) Å. The shortest and the longest coordination distances correspond to the Ir–C and Ir–O bonds, respectively. The Ir–O bond distance in 5 (2.155(3) Å), which is considerably longer than the mean value of 2.088 Å reported, reflects the pronounced *trans* influence of the aromatic rings. [10]

2.3. Computational Studies

To study the electronic structures and understand the nature of excited states of these multifunctional iridium phosphors, density functional theory (DFT) and time-dependent DFT (TDDFT) calculations were carried out using B3LYP hybrid functional theory for 1, 3 and 5 based on their geometries obtained from X-ray diffraction studies. From the TDDFT calculations, we note that the $S_0 \rightarrow S_1$ transitions correspond to the HOMO→LUMO transitions with non-zero oscillator strengths (table 1). Therefore, we simply focus on the HOMO-LUMO analyses in the following discussion. The results of DFT calculations show that they all have similar HOMO and LUMO levels in terms of their orbital characters (Figure 2). The contour plots show that the HOMOs mainly consist of the "t_{2g}" d orbitals from the metal center mixing with the π orbitals of the phenyl rings that are directly bonded to the metal center. The LUMOs correspond to the lowest π orbital of the conjugated organic ligands, to which the N-containing rings contribute more. It is clear that the aryl groups essentially do not contribute to the HOMOs of 3 and 5. Hence, the HOMO stability and the emission energy gap are not governed by the identity and inductive influence of X. Substituent effect on the LUMO levels is also expected to be very small.

Table 1. Percentage contribution of the metal dπ orbitals to HOMO and LUMO levels together with the TDDFT calculation results

Complex	Contribution of d orbitals to HOMO	Contribution of d orbitals to LUMO	The largest coefficient in the CI expansion of the S_1 state [a]	The oscillator strength (*f*) of the $S_0 \rightarrow S_1$ transition
1	24.5 %	3.6 %	H→L: 0.53	0.018
3	36.2 %	1.5 %	H→L: 0.69	0.021
5	36.3 %	2.6 %	H→L: 0.69	0.025

[a]H→L represents the HOMO to LUMO transition. CI stands for configuration interaction.

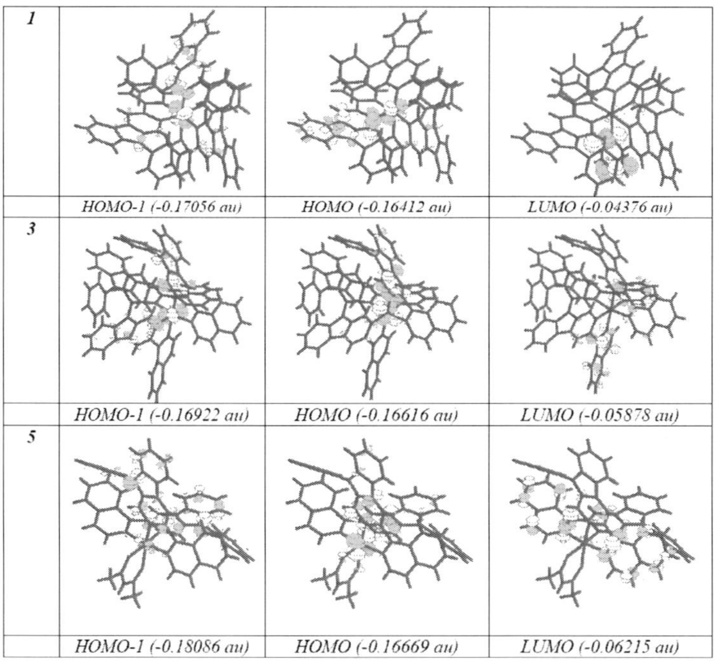

Figure 2. Contour plots of dominant excitation orbitals of 1, 3 and 5 (1 au = 27.2114 eV).

2.4. Thermal Stability and Photophysical Characterization

The thermal stability of 1–10 was examined by thermal gravimetric analysis (TGA) under nitrogen stream (table 2). All of them show excellent thermal stability, with their onset decomposition temperatures (T_{decomp}) of 345–502 °C (table 1) and in general the homoleptic species is thermally more stable than the corresponding heteroleptic congener. In addition, compounds 1–6 can be evaporated under vacuum without decomposition. These desirable features would be beneficial to the long-term stability of evaporated OLED devices fabricated from these materials. The TGA data also suggest that the presence of rigid carbazole unit can enhance the thermal stability relative to their parent complexes with simple ppy and piq groups. However, because of the high molecular weights of 7–10, they do not sublime readily under vacuum and should preferrably be used as solution-processed OLED devices (vide infra).

Table 2. Photophysical and thermal data for 1–10

	Absorption (293 K)	Emission (293 K)					Emission (77 K)		T_{dec} [°C]
	λ_{abs} [nm] [a] CH_2Cl_2	λ_{em} [nm] CH_2Cl_2	ΦP [b]	τP [μs] [c]	kr [s^{-1}]	$k$nr [s^{-1}]	λ_{em} [nm] CH_2Cl_2	τP [μs] [c]	
1	291sh (4.69) 316 (4.69) 398sh (0.20)	515	0.43	0.46	9.3×10^5	1.2×10^6	493 525sh	25.8	477
2	281sh (4.89) 315 (4.85) 400sh (0.16)	514	0.33	0.54	6.1×10^5	1.2×10^6	505	33.9	421
3	234 (9.08) 268 (5.48) 327 (4.54) 388 (2.73) 450sh (0.49)	620	0.19	0.68	2.8×10^5	1.2×10^6	608 658sh	5.41	432
4	234 (10.19) 267 (6.80) 328 (5.33) 393 (3.17) 455sh (1.41)	624	0.16	0.72	2.2×10^5	1.1×10^6	617 667sh	9.11	433
5	232 (12.65) 331 (6.63) 363 (4.96) 399 (3.85) 499sh (0.13)	628	0.30	0.79	3.8×10^5	8.9×10^5	626 678sh	6.31	398
6	232 (6.17) 265 (4.30) 321 (3.07) 364 (2.09) 405 (1.74) 500sh (0.21)	628	0.27	0.82	3.3×10^5	8.9×10^5	619 673sh	5.92	345
7	294 (14.8) 342 (18.2) 398 (5.40) 463 (1.50)	552 593sh	0.20	0.09	2.2×10^6	8.9×10^6	552 599	10.45	478
8	293 (6.70) 347 (10.4) 466 (0.90)	562 605sh	0.12	0.09	1.3×10^6	9.8×10^6	561 610	12.53	502
9	294 (4.40) 347 (5.00) 466 (0.4)	560 604sh	0.10	0.12	8.3×10^5	7.5×10^6	556 601	10.66	354
10	295 (10.9) 366 (14.0) 466 (1.50)	566 608sh	0.09	0.13	6.9×10^5	7.0×10^6	565 612	15.08	360

[a] ε values (10^4 M^{-1} cm^{-1}) are shown in parentheses.

[b] Measured in degassed CH_2Cl_2 relative to *fac*-[Ir(ppy)$_3$] ($\Phi_P = 0.40$), $\lambda_{ex} = 400$ nm.

[c] In degassed CH_2Cl_2 at 293 K. sh = shoulder.

Absorption and photoluminescence (PL) spectra of 1–6 were measured in CH_2Cl_2 solutions (table 2). Figures 3 and 4 show a comparison of their UV-Vis and PL spectra measured at 293 K. For all of them, the higher energy absorption peaks below 400 nm are assigned to the spin-allowed $^1\pi-\pi^*$ transitions of the cyclometalated ligands. Common to other Ir(III) complexes, [1-3] there is a weak but well-resolved peak above 400 nm with lower extinction coefficients which can be originated from a spin-forbidden, ligand-centered $^3\pi-\pi^*$ transition facilitated by spin-orbit coupling, together with an admixture of some ^{1}MLCT and ^{3}MLCT states (MLCT = metal to ligand charge transfer). The nature of the X group on carbazole does not have a great influence on the absorption and PL spectra of 1–6. This can also be explicable from the DFT data (vide supra). Both 1 and 2 emit strong green phosphorescence (λ_{em} ~ 514 nm, Stokes shifts ~ 200 nm) at room temperature in CH_2Cl_2. Their triplet energy levels (T_1 ~ 2.41 eV) are close to each other and that of *fac*-[Ir(ppy)$_3$] (2.43 eV). [1a] Phosphors 3–6 emit deep red PL bands (λ_{em} in CH_2Cl_2 = 620, 624, 628, 628 nm for 3–6, respectively) and their spectra are broad and structureless implying that the lowest triplet states of these complexes are likely dominated by ^{3}MLCT excited states. Both intense ligand-based and weaker MLCT absorptions are red-shifted from 3 to 5 (or 4 to 6). The slightly red-shifted emissions observed for acac-based complexes may be associated with the comparatively higher ligand field of the acac anion.

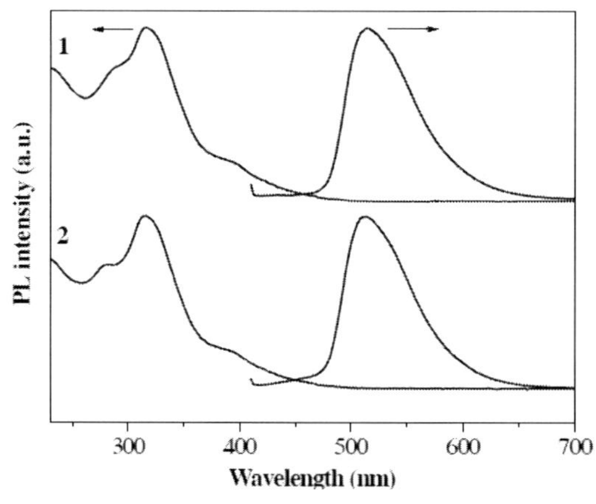

Figure 3. Optical absorption and PL spectra of green-emitting 1 and 2 in CH_2Cl_2 at 293 K.

All of the complexes are also intensely luminescent in the glass matrices at 77 K and undergo notable rigidochromic blue shifts upon cooling of the solutions to 77 K. The hypsochromic shifts observed at 77 K are mainly due to the solvent reorganization in a fluid solution at low temperature that can stabilize the charge transfer states prior to emission. [11] This process is significantly impeded in a rigid matrix at 77 K, and hence phosphorescence occurs at a higher energy. For instance, at low temperatures, the vibronic fine structures in the PL spectra of 3–6 become more apparent (Figure 4). We further probed the photophysical properties of the complexes by measuring their phosphorescence quantum yields (Φ_P) in dilute degassed CH_2Cl_2 solutions at 293 K (table 2). The Φ_P values of 1 and 2 in CH_2Cl_2 are ca. 0.33–0.43, relative to a *fac*-[Ir(ppy)$_3$] standard (Φ_P = 0.40). [12] We note that 5 has the highest Φ_P of 30 %, which is larger than 3 (19 %). The Φ_P of Ir complexes with X = H are slightly higher than those with X = OMe.

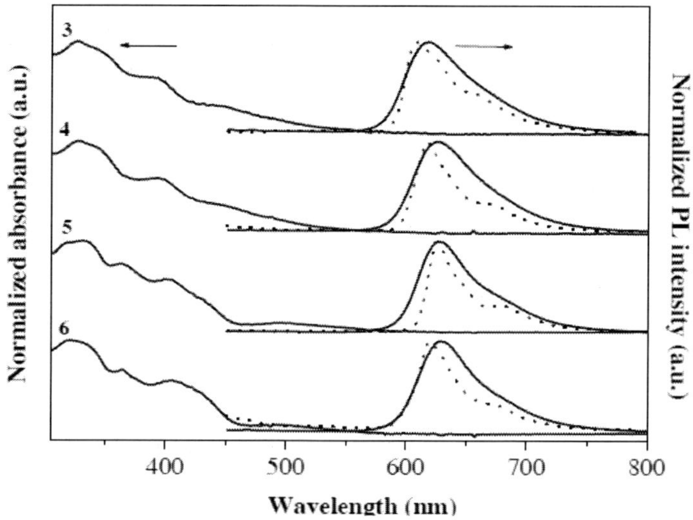

Figure 4. Optical absorption and PL spectra of red-emitting 3–6 in CH_2Cl_2 at 293 K (conc. ~ 10^{-5} M). The PL spectra at 77 K are also shown as dotted lines.

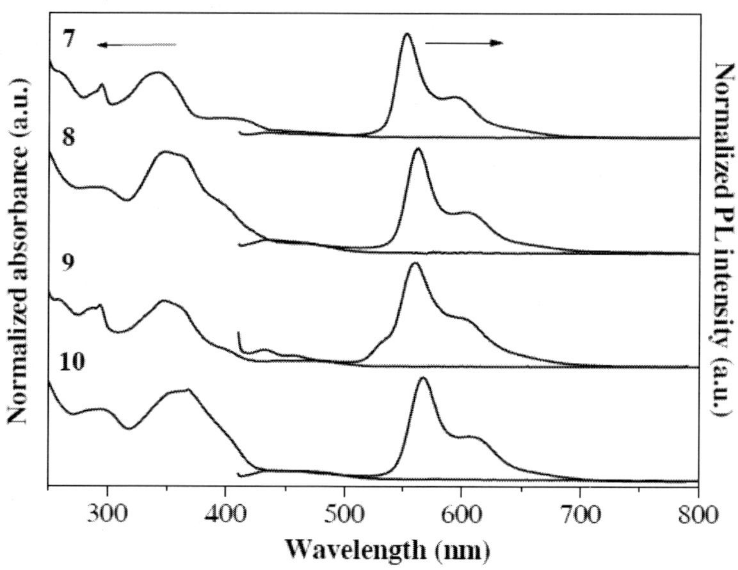

Figure 5. Optical absorption and PL spectra of orange-emitting 7–10 in CH_2Cl_2 at 293 K.

The photophysical properties of 7–10 were investigated in CH_2Cl_2 (table 2). Their absorption and PL spectra at 293 K are shown in Figure 5. The intense peaks in the short wavelength region below 400 nm with extinction coefficients (ε) in the order of 10^5 are attributed to the spin-allowed $^1\pi$–π^* transitions of the aromatic segments which closely resemble the spectra of the free ligands L_5 (λ_{abs} = 294, 338 nm) and L_6 (λ_{abs} = 304, 344 nm). The weaker absorption tails that appear above 400 nm with much smaller ε are likely due to charge-transfer transitions. [13] These data suggest that there is a substantial mixing of ligand-based $^3\pi-\pi^*$ states, spin forbidden ^{3}MLCT and higher-lying ^{1}MLCT transitions induced by the spin-orbit coupling effect. [14] The spin-orbit coupling is enhanced by the presence of closely-spaced $\pi-\pi^*$ and MLCT states and the heavy-atom effect of Ir center. [15] Under photoexcitation, they emit an intense orange emission.

By comparing with the emission spectra of the free ligands L_5 and L_6 which emit blue fluorescence at 406 and 414 nm, respectively, each of 7–10 shows a strong room-temperature phosphorescence (λ_{em} in CH_2Cl_2 = 552, 562, 560, 566 nm for 7–10, respectively) from the predominantly ligand-centered $^3\pi-\pi^*$ excited state which is characterized by a large Stokes shift (> 100 nm) from the lowest energy absorption peak. The vibronic progression of ca. 1222–1302 cm^{-1} (ν_{0-1}) in the emission profile corresponds to the aromatic stretching of the cyclometalated ligands which is diagnostic of a substantial involvement of the ligand-centered $\pi-\pi^*$ transitions in the emission. We also anticipate that there is a weakening of the spin-orbit coupling and hence an increasing amount of the $^3\pi-\pi^*$ character in the lowest energy excited state as the size of the cyclometalating organic group increases, consistent with some previously reported data. [16] The emission spectra of 7–10 show the vibrational fine structures at low temperatures. Such vibronic structures plus the molecular size factor suggest an insignificant contribution from the ^{3}MLCT states for 7–10. A slight red shift of the emission wavelength was observed from 7 to 8 and 9 to 10, which can be attributed to the longer effective conjugation length in L_6 than L_5 because of the increased electron-donating ability of the carbazole unit of the former case. The higher ligand-field strength of acac anion also leads to the bathochromic shift in λ_{em} for 9 and 10 relative to 7 and 8, respectively. As compared to the non-carbazole complex [Ir(Flpy)$_3$] (HFlpy = 2-(9,9-diethylfluoren-2-yl)pyridine) [17] (λ_{abs} = 321, 336, 405 nm, λ_{em} = 545 nm), phosphors 7 (λ_{abs} = 294, 342, 398, 463 nm, λ_{em} = 552 nm in CH_2Cl_2) and 8 (λ_{abs} = 293, 347, 466 nm, λ_{em} = 562 nm) show an obvious red shift in the absorption and emission features in the presence of the electron-rich carbazole moiety which can destabilize the ground state of the complex by electron donation. The introduction of carbazole derivatives extends the electronic π-conjugation framework of the ligands which can decrease the LUMO energy levels.

We further probed the photophysical properties of 7–10 by measuring their Φ_P in CH_2Cl_2 solutions at room temperature. Among these compounds, we found that 7 has the highest Φ_P of 20 %, which is higher by about two times than its heteroleptic analogue 9 (10 %). A similar case was observed for the complexes derived from ligand L_6. The short τ_P values together with the relatively high Φ_P should be advantageous to fabricating highly efficient OLED devices.

The room temperature lifetimes (τ_P) of 1–10 fall in the sub-microsecond range (τ_P = 0.46–0.82 μs for 1–6 and 0.08–0.13 μs for 7–10 respectively), which coupled with the large Stokes shift of the PL peak confirm the phosphorescent excited states in each case. Their sensitivity towards oxygen is also consistent with this phenomenon. [18] It is known that the phosphorescence lifetime is a key factor that causes triplet-triplet annihilation in the OLED operation. The longer τ_P of the material usually causes a more significant triplet-triplet annihilation. [19] The short lifetimes of the complexes can increase spin state mixing and reduce the chance of device efficiency decay. The τ_P values at 77 K are longer than those at 293 K. The emission lifetimes of the complexes are generally determined by nonradiative decay rates which usually decrease with lowering temperature. Thus, a lengthening of the lifetime at low temperatures would be anticipated. Assuming that the emitting state of a heavy metal phosphor is formed with unit efficiency, one can calculate the radiative (k_r) and nonradiative (k_{nr}) rate constants using the relationships $k_r = \Phi_P/\tau_P$ and $k_{nr} = (1 - \Phi_P)/\tau_P$. [20] The calculated k_r and k_{nr} values are listed in table 2. Complexes 1–6 have a similar order of magnitude in the k_r values, however, the k_{nr} values are about one order of magnitude faster than those for k_r. Likewise, phosphors 7–10 have reasonably similar k_{nr} values, while the k_r values for 7 and 8 are approximately one order of magnitude faster than 9 and 10. Complexes 7 and 8 (or 9 and 10) have similar k_r values, as can be expected from the comparable quantum efficiencies and lifetimes for both series of compounds. Also, it is found that the k_r values for 7 and 8 ($1.3–2.2 \times 10^6$ s^{-1}) are larger than that found for [Ir(Flpy)$_3$] (2.5×10^5 s^{-1}), and this would be advantageous to the design of highly efficient devices based on light-energy harvesting from the triplet excitons.

2.5. Redox and Electronic Characterization

The cyclic voltammograms of the complexes were measured in THF containing 0.1 M [NBu_4]PF_6 using ferrocene as the internal standard. On the basis of the oxidation and reduction potentials, the HOMO and LUMO energy levels of 1–10 were estimated with reference to the energy level of ferrocene (4.8 eV below the vacuum level) and are listed in table 3. [21] During the anodic scan, all of the complexes show one reversible wave derived from the Ir(III)/Ir(IV) oxidation. Cyclic voltammetry shows that each of 1 and 2 was oxidized at less-positive potentials (0.15, 1; 0.16 V, 2) than *fac*-[Ir(ppy)$_3$], indicating a greater propensity for 1 and 2 to lose an electron. Cathodic sweeps in THF reveal an irreversible wave at –2.91 to –2.84 V. Hence, the HOMO levels are significantly elevated (–4.95, 1; –4.96 eV, 2) relative to *fac*-[Ir(ppy)$_3$] (–5.05 eV). This indicates that they have a lower ionization potential (IP) than *fac*-[Ir(ppy)$_3$], leading to a better hole injection (HI) and HT ability in 1 and 2 to favor the transport of holes. Our complexes show T_1 levels well below the level of CBP host (2.56 eV) to prevent reverse energy transfer from the guest back to the host, and to confine triplet excitons on guest molecules.

Consistent with our DFT results, changes in the substituents on the aryl group of the ligand do not affect markedly the HOMO level of the complexes. The oxidation potentials for 3 and 4 (0.16 and 0.14 V, respectively) or 5 and 6 (0.24 and 0.21 V, respectively) are similar. A stabilized HOMO level was observed in the acac-based complexes 5 and 6 due to a stronger ligand field strength by the ancillary ligand. By comparing with the parent compounds [Ir(piq)$_3$] and [Ir(piq)$_2$(acac)] (Hpiq = 1-phenylisoquinoline), [22] replacement of the phenyl ring in Hpiq with 9-arylcarbazole group can largely tune the HOMO level. The higher-lying HOMO levels of 3 and 4 (–4.96 and –4.94 eV, respectively) as well as 5 and 6 (–5.04 and –5.01 eV, respectively) are notably shifted to higher energy levels relative to [Ir(piq)$_3$] (–5.11 eV)[5h] and [Ir(piq)$_2$(acac)] (–5.32 eV), [22] indicating a greater ease of oxidation for these complexes.

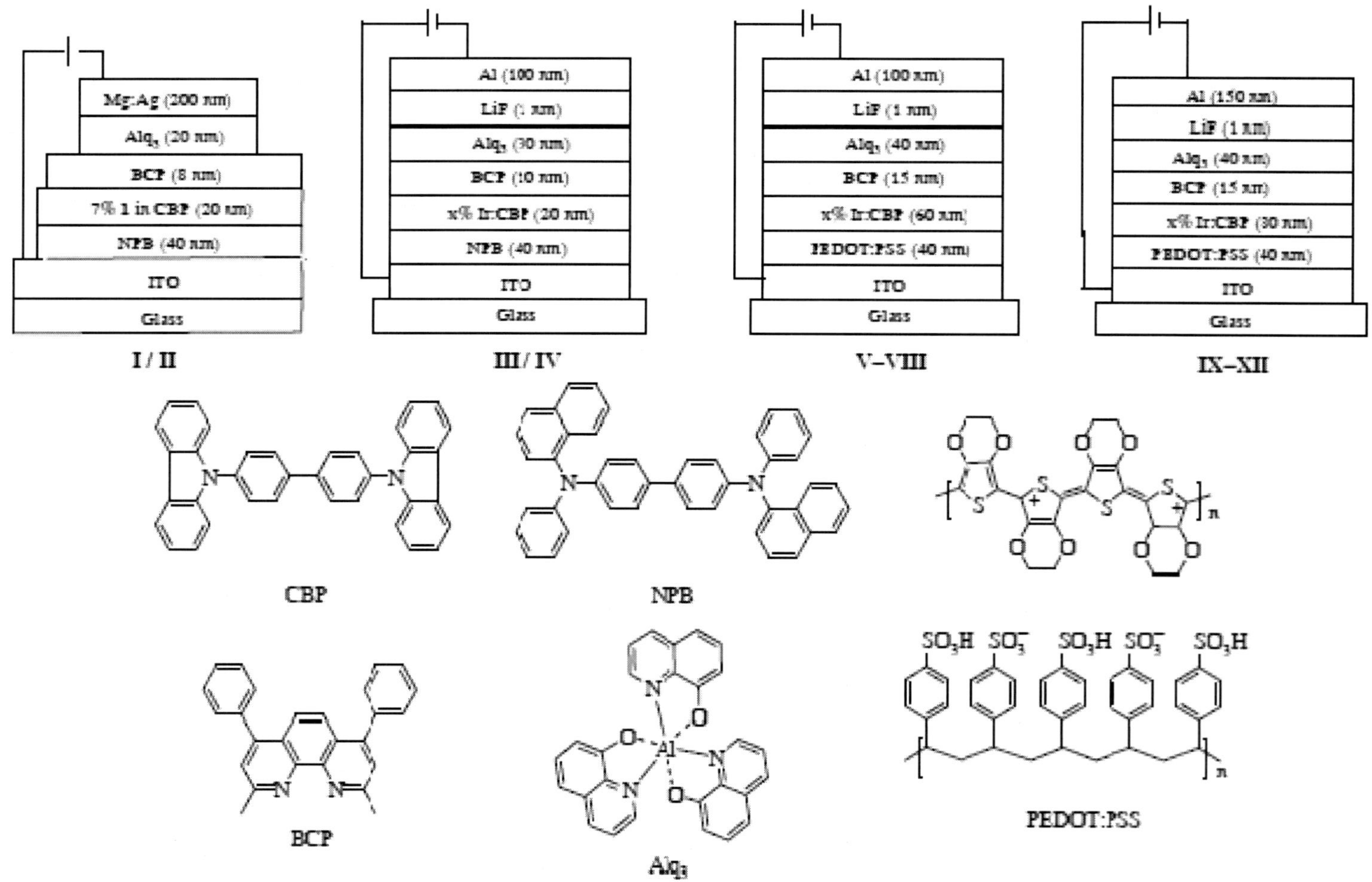

Figure 6. The general structures for OLED devices I–XII and the molecular structures of the relevant compounds used in these devices.

Table 3. Electrochemical properties and frontier orbital energy levels of 1–10

Complex	HOMO [eV]	LUMO [eV]
1	–4.95	−1.89
2	–4.96	−1.96
3	–4.96	−2.43
4	–4.94	−2.43
5	–5.04	−2.49
6	–5.01	−2.49
7	–5.04	−2.65
8	–4.96	−2.54
9	–5.09	−2.72
10	–5.06	−2.67

This indicates that addition of the electron-rich carbazole moiety can effectively rise the HOMO level and therefore complexes 3–6 are expected to be better HT materials than [Ir(piq)3] and [Ir(piq)2(acac)]. Cathodic sweeps in THF reveal a reversible wave within –2.31 to –2.37 V for 3–6.

Each of 7–10 exhibits a Ir(III)/Ir(IV) reversible oxidation with the redox potential in the region of 0.16–0.29 V in THF. These oxidation potentials are lower for the homoleptic complexes 7 and 8 than their corresponding heteroleptic complexes 9 and 10. Notably, the incorporation of 9-phenylcarbazole groups to the fluorene core caused a negative shift in the anodic $E_{1/2}$ by ca. 70 mV as observed by switching from the unsubstituted [Ir(Flpy)$_3$] (0.23 V) to 8, indicating a greater ease for 8 to lose an electron. We find that the HOMO levels are higher for the 3-substituted 9-phenylcarbazole end group than for the *N*-coordinated carbazole unit (i.e. 8 > 7 and 10 > 9). This can be ascribed to the more effective electronic conjugation in L_6 which can destabilize the HOMO, leading to reduced oxidation potentials. The energy gap becomes slightly lower for the acac complexes 9 and 10, which is consistent with the fact that the auxiliary ligand with a stronger ligand-field strength could stabilize the HOMO level. Saliently, the HOMO value of 8 is elevated to −4.96 eV relative to the vacuum level after incorporation of 9-phenylcarbazole when compared to [Ir(Flpy)$_3$] (−5.02 eV). This indicates that complex 8 is a better HT material than its non-carbazole capped analogue. The HOMO levels of 7–10 are relatively close to the Fermi level of PEDOT (–5.0 eV), [23] resulting in a good HI contact and an improvement in the hole carrier balance in the OLEDs even without the need for an additional HT layer (vide infra).

2.6. Electrophosphorescent OLED Fabrication and Characterization

2.6.1. Vacuum-Evaporated Green and Red OLEDs

Figure 6 depicts the architectures used for the fabrication of various OLED devices in our investigations. Highly efficient green-emitting OLEDs I and II were fabricated using 1 and 2, respectively, as phosphorescent dopant emitters in a four-layer structure of ITO/NPB (40 nm)/6% 1 or 2 in CBP (20 nm)/BCP (8 nm)/Alq_3 (20 nm)/Mg:Ag (200 nm) (ITO = indium tin oxide). CBP acts as the host material for the electrophosphor and 4,4'-bis[*N*-(1-naphthyl)-*N*-phenylamino]biphenyl (NPB) as a hole-transport layer (HTL). A thin hole-blocking barrier layer of 2,9-dimethyl-4,7-diphenyl-1,10-phenanthroline (BCP) inserted between the CBP and the electron-transporting layer (ETL) of tris [8-hydroxyquinoline] (Alq_3) was necessary to confine excitons within the luminescent zone and hence maintain high efficiencies. The electroluminescence (EL) spectra consist only of green phosphor emission without any residual emission from the host and/or adjacent layers even being operated at high drive currents, which indicates complete energy transfer from the host exciton to the phosphor upon electrical excitation. The EL spectra for the devices, which closely resemble the PL spectra, exhibited no significant change with variation of current and operating bias voltages from 5 to 12 V and each of the devices I and II shows a sharp EL peak at 508 nm (Figure 7). The turn-on voltages (defined as the bias at a brightness of 1 cd m^{-2}) are 4.4 and 4.8 V for devices I and II, respectively, with the Commission Internationale de L'Eclairage (CIE) green color coordinates being (0.24, 0.63) and (0.27, 0.60).

Phosphors 1 and 2 show outstanding EL performance, particularly thanks to the short τ_P and good HT properties of carbazole, which may help diminish the quenching of the triplet exciton. Complex 1 was shown to be a more efficient electrophosphor than 2, suggesting that the F substituents are not critical here in relieving intermolecular interactions in the complex. Moreover, since 2 has a triplet lifetime comparable to that of 1, its poor EL performance may be attributed to its lower Φ_P relative to 1. The performance parameters of device I (maximum external quantum efficiency (η_{ext}) = 11.6 % photons/electron, luminance efficiency (η_L) = 38 cd A^{-1}, power efficiency (η_P) = 24 lm W^{-1}) obviously outweigh those of prototypical green-light-emitting device of ITO/NPB/*fac*-[Ir(ppy)$_3$]:CBP/BCP/Alq_3/Mg:Ag (maximum η_{ext} = 7.5 %, η_L = 26 cd A^{-1}, η_P = 19 lm W^{-1}) under similar device structure and this represents a tremendous enhancement of ~ 55 % in η_{ext}. [1a] The maximum luminance (L) of device I is 14730 cd m^{-2} at 12 V, with the full width at half maximum (fwhm) of only 56 nm at 8 V. With 2 as the dopant, device II was demonstrated to show a maximum L of 19360 cd m^{-2} and η_L of ~ 22 cd A^{-1}, corresponding to η_{ext} and η_P of 6.7 % and 13 lm W^{-1}.

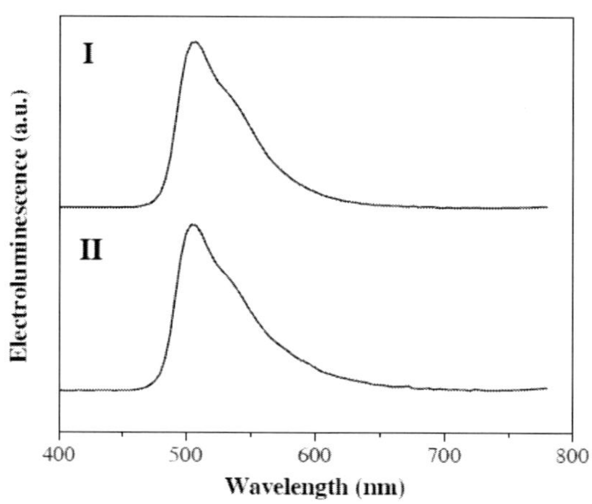

Figure 7. EL spectra of OLED devices I and II.

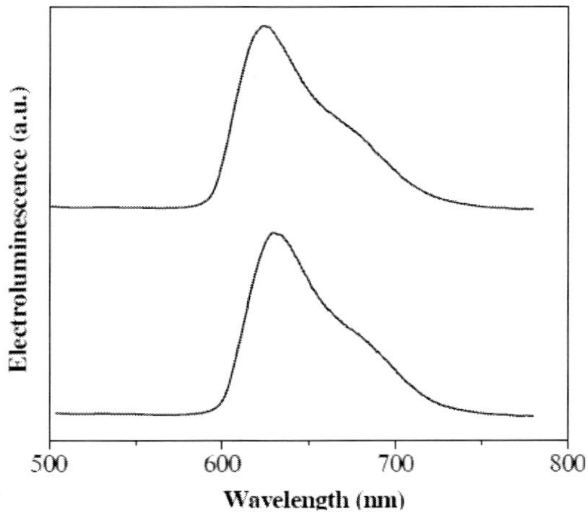

Figure 8. EL spectra of OLED devices (a) IIIa and (b) IVa.

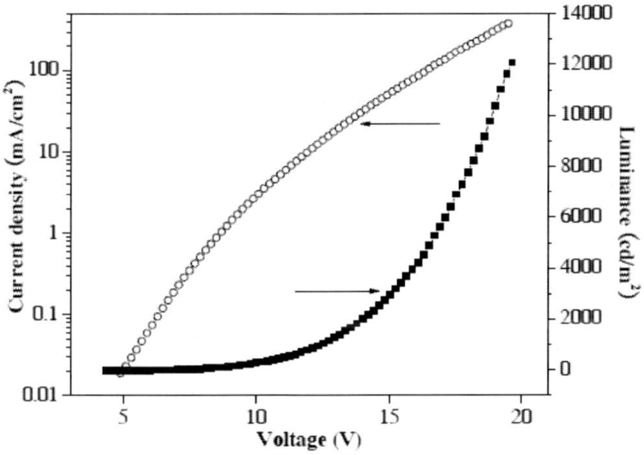

Figure 9. *I–V–L* characteristics of the electrophosphorescent OLED device IVc.

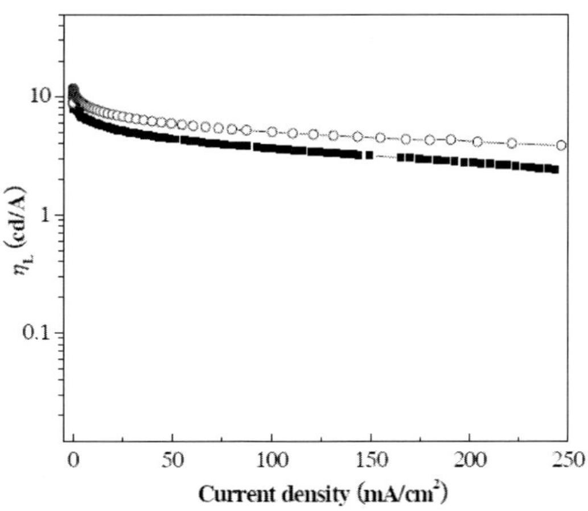

Figure 10. Luminance efficiency as a function of current density for devices (a) IIIa (■) and (b) IVa (○).

To date, there is a greater success in the research of a green triplet emitter than that of a red one. Many red organic dyes currently in use do not show a good compromise between OLED efficiency and color purity, which represents a dilemma facing dopant-based red OLEDs. In other words, efficient and bright dopants are not red enough, and red-enough dopants are not efficient and bright. Highly efficient OLEDs exhibiting saturated red EL with CIE coordinates at (0.67, 0.33) have attracted increasing attention for their applications in full-color flat-panel displays. [24] Optimization of the OLED efficiency/color purity trade-off is thus a key issue for the realization of highly efficient phosphors showing good color purity of red EL. Among the red-emitting iridium complexes, [Ir(piq)$_3$] is one of the most important one. [17,25] To improve the charge injection and transport in [Ir(piq)$_3$], the incorporation of bulky and HT carbazole group as a core part in the ligand frame in the Ir isoquinoline complexes is anticipated to help in reducing the barrier height for hole injection and decreasing the triplet-triplet annihilation. Representative multilayer EL devices III and IV were constructed using dopants 3 and 5. LiF (1 nm) acts as the electron injection layer and Al as the cathode (Figure 6). Device V was fabricated using 3 as the dopant with OLED structure: ITO/NPB (60 nm)/x % 3:CBP (30 nm)/BPhen (20 nm)/LiF:Al (1:60 nm) (BPhen: 4,7-diphenyl-1,10-phenanthroline). The configuration of device VI is ITO/NPB (60 nm)/x% **3**:CBP (25 nm)/TPBI (30 nm)/LiF:Al (1:120 nm) (TPBI: 1,3,5-tris[*N*-(phenyl) benzimidazole]benzene). We used CBP as the host materials because of its proven good performance as host for iridium complexes and theoretical confirmation of favorable triplet energy. [1b,26] Different dopant concentrations were used to get the optimum weight percentages for devices with the highest efficiencies. All of the complexes display voltage-independent red EL peaks at 620–628 nm, which matched well the PL spectra in solution. The close resemblance between the PL and EL spectra indicates the absence of aggregation or π-stacking at these doping levels and the origin of EL emission from the triplet excited states was justified. Figure 8 depicts the EL spectra of IIIa and IVa (both at 4 wt.-% level), and the current–voltage–luminance (*I–V–L*) characteristics of device IVc (at 10 wt.-% level) is shown in Figure 9. There are no undesirable emission peaks from CBP or Alq_3 in each case, signaling effective energy transfer from the CBP host to the Ir(III) dopant in the emissive layer upon electrical excitation. The EL characteristics of devices III–VI are summarized in table 4. The maximum external quantum efficiency (η_{ext}) of devices IIIa and IVa can reach 11 and 12 % photons per electron at 4 wt.-% doping concentration with CIE color coordinates of (0.66, 0.34) and (0.68, 0.33) for 3 and 4, respectively. To our knowledge, these η_{ext} values are some of the highest for saturated red OLEDs. Their performance compares well with the efficiency for phosphorescent polymer LEDs with saturated red emission (~ 12 %) [27] and is much better than those for first-generation red-emitting Ir dendrimers based on meta-bonded phenylene (5.70 and 4.25 %). [28] Devices III and IV also outperform the best vacuum-evaporated device with the dopants [Ir(piq)$_3$] (η_{ext} ~ 10.3 % and CIE (0.68, 0.32)) [17] and [Ir(PEQ)$_2$(acac)] (PEQ = 2-((*E*)-2-phenyl-1-ethenyl)quinoline, η_{ext} ~ 7.2 % and CIE (0.66, 0.33)) in CBP host. [29] We attribute this excellent performance to the hole-trapping carbazolyl structure of 3 and 5 in enhancing the HI/HT properties. The CIE values are close to the National Television Standards Committee recommended red (standard red specification) at x = 0.670 and y = 0.330 for a video display. [30] The corresponding peak η_L is 11.00 cd A^{-1} for IIIa and 10.15 cd A^{-1} for IVa (see Figure 10). It is worth noting that device IV appears better in performance than those of prototypical device ITO/NPB/[Ir(piq)$_2$

(acac)]:CBP/BCP/Alq_3/Mg:Ag (maximum η_{ext} = 7.0 %, η_L = 6.08 cd A^{-1}, η_P = 1.88 lm W^{-1}) under similar device structure. [31]

Table 4. Performance of electrophosphorescent OLEDs based on vacuum-deposited emissive layer

Device	Dopant [wt.-%]	$V_{turn-on}$ [V]	L_{max} [cd m^{-2}] [c]	η_{ext} [%]	η_L [cd A^{-1}]	η_P [lm W^{-1}]	λ_{max} [nm] [d]
I	1 (6)	4.4	14730 (12)	6.39 [a] 4.56 [b] 11.56 (5.2) [c]	21.06 14.90 38.01 (5.2)	5.16 3.25 23.92 (4.8)	508 (0.24, 0.63)
II	2 (6)	4.8	19360 (12)	4.50 3.44 6.69 (5.6)	14.44 11.03 21.35 (5.6)	6.75 3.91 12.33 (5.2)	508 (0.27, 0.60)
IIIa	3 (4)	4.6	5846 (24.7)	8.27 [a] 5.42 [b] 10.87 (5.5) [c]	8.14 5.52 11.00 (5.5)	2.92 1.35 6.16 (5.3)	624 (0.66, 0.34)
IIIb	3 (6)	5.1	8825 (20.9)	7.59 5.44 8.55 (5.9)	7.66 5.48 8.86 (6.5)	2.64 1.33 4.61 (5.9)	624 (0.66, 0.34)
IIIc	3 (8)	4.8	8620 (20.9)	7.66 5.91 7.75 (5.7)	7.74 5.97 7.84 (5.7)	2.95 1.60 3.43 (6.5)	624 (0.66, 0.34)
IIId	3 (10)	3.4	8744 (23.7)	7.18 5.93 7.24 (7.5)	7.27 5.99 7.33 (7.5)	2.86 1.65 3.43 (6.5)	624 (0.66, 0.34)
IVa	5 (4)	4.2	10750 (19.5)	9.78 7.37 11.76 (6.3)	8.45 6.36 10.15 (6.3)	2.86 1.60 5.25 (5.1)	628 (0.68, 0.33)
IVb	5 (6)	4.7	9971 (22.3)	10.25 7.43 11.70 (6.7)	8.88 6.46 10.10 (6.7)	2.86 1.46 5.38 (5.7)	628 (0.68, 0.33)
IVc	5 (8)	4.9	11776 (19.9)	10.23 7.71 11.13 (7.1)	8.81 6.63 9.60 (7.1)	2.97 1.60 5.10 (5.7)	628 (0.68, 0.33)
IVd	5 (10)	5.4	12510 (19.7)	9.97 7.99 10.34 (7.7)	8.62 6.86 8.93 (7.7)	2.99 1.71 4.26 (5.5)	628 (0.68, 0.33)

[a] Values collected at 20 mA cm^{-2}.
[b] Values collected at 100 mA cm^{-2}.
[c] Maximum values of the devices. Values in parentheses are the voltages at which they were obtained.
[d] CIE coordinates.
[x, y] in parentheses.

We can surmise that it is probably the carbazolyl structural moiety that leads to a more balanced electron and hole recombination in the CBP host, and/or that the rigid carbazole spacer plays a crucial role in alleviating the self quenching of the luminophores, and consequently increase the emission efficiency. Such a good performance can be attributed to the relatively short emission lifetimes of our materials. Typical efficiency roll-off at higher currents, caused by a combination of triplet-triplet annihilation and field-induced quenching effects, is also observed here in devices I–IV, yet at the benchmark luminance of 100 cd m^{-2}, the efficiencies remain high for most of these devices. The lower luminosity of the red-emitting devices III–VIII is indeed not surprising and should be due to their characteristic pure red emission in a spectral region where the eye has poor sensitivity. Though the performance of these devices has yet to be optimized, it is our intention to highlight the potential merit of this prominent class of carbazole-based Ir phosphors in high-efficiency OLED applications.

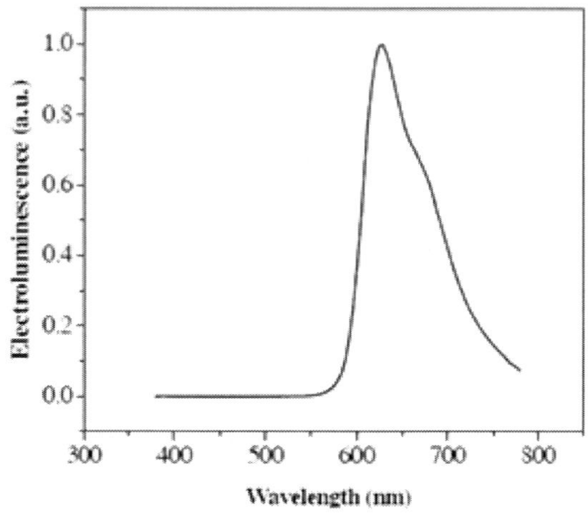

Figure 11. EL spectra of the OLED device VIc.

2.6.2. Solution-Processed Red and Orange OLEDs

Recently, the great demand for the fabrication of large area and flexible displays stimulates a great deal of research attention on solution-processed OLED materials. To demonstrate the potential of our materials in this area, we have used phosphors 36 for the fabrication of solution-processed red-light OLEDs and investigated their EL performance by using the relatively cheaper solution-processing technique for the preparation of emissive layer. The good film-forming properties of these Ir phosphors can ensure a good chemical compatibility with the CBP host which leads to a homogeneous distribution of the Ir dopant in CBP. The devices V–VIII consist of the multilayer configuration: ITO/PEDOT:PSS (40 nm)/x% Ir:CBP (60 nm)/BCP (15 nm)/Alq_3 (40 nm)/LiF:Al (1:100 nm) for dopants 3–6, respectively with x between 3 to 15 wt.-%. Poly(3,4-ethylenedioxythiophene):poly(styrene sulfonate) (PEDOT:PSS) only acts as the HI layer but the conventional HTL of NPB was not necessary for decent device performance. While [$Ir(piq)_3$] does not dissolve well in many common solvents such as *p*-xylene, chlorobenzene and toluene, our carbazole-derived Ir(III) complexes 3–6 are readily soluble in these solvents and amenable to spin-coating fabrication. Each of the devices exhibits a strong red EL peak in the narrow wavelength range of 624–628 nm for 3–6 and a representative example is depicted in Figure 11 for device VIc. The relatively high driving voltages as compared to the vacuum-evaporated devices may be due to the thicker emissive layer of CBP:Ir dopant used and the large hole barrier at the PEDOT:PSS/CBP interface (HOMOs of CBP ~ 6.2 eV [32] versus PEDOT ~ 5.0 eV). We note that the EL peaks are almost invariant of the driving voltage and also do not show any concentration dependence. The EL performance is still impressive even for solution-processed devices, presumably because the carbazolyl groups in 3–6 are able to play a vital role in HT. The best EL efficiencies were observed in device VIa with maximum η_{ext} of 7.65 %, η_L of 6.18 cd A^{-1} and η_p of 1.07 lm W^{-1} at a concentration of 5 wt.-%. The devices V, VII and VIII show the corresponding peak efficiencies at $\eta_{ext} = 5.60$ %, $\eta_L = 4.84$ cd A^{-1} and $\eta_p = 2.41$ lm W^{-1} for Vc at 10 wt.-% concentration, $\eta_{ext} = 7.18$ %, $\eta_L = 4.65$ cd A^{-1} and $\eta_p = 0.67$ lm W^{-1} for VIIb at 5 wt.-% concentration, and $\eta_{ext} = 6.24$ %, $\eta_L = 3.80$ cd A^{-1} and $\eta_p = 0.63$ lm W^{-1} for VIIIa at 3 wt.-% concentration. Figure 12 illustrates the *I–V–L* curves for the representative device VIc. Figure 13 shows η_L as a function of current density for device VIIc and VIIIc. Devices suffered less from a decay in efficiency at an elevated current density in many occasions. For instance, for devices Vc, Vd, VId and VIId, the decrease is not significant at all which only contributes to a loss of less than 5.0, 9.6, 2.3 and 5.7 %, respectively, in η_{ext} when the current density increases from 20 to 100 mA cm^{-2}. The short triplet state lifetime of our red phosphors should be advantageous to the realization of highly efficient OLEDs by minimizing the T–T annihilation effectively. The performance of these solution-processed devices is superior to that of the similarly fabricated red-emitting Ir dendrimer (~ 5.7 %). [28] Gratifyingly, the CIE colors of (0.67, 0.33) and (0.68, 0.32) for V–VIII correspond well to the true red region which were shown to be independent of the current density.

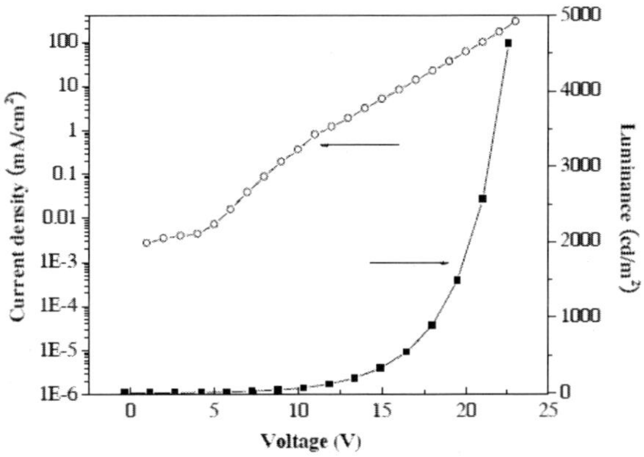

Figure 12. *I–V–L* characteristics of the electrophosphorescent OLED device VIc.

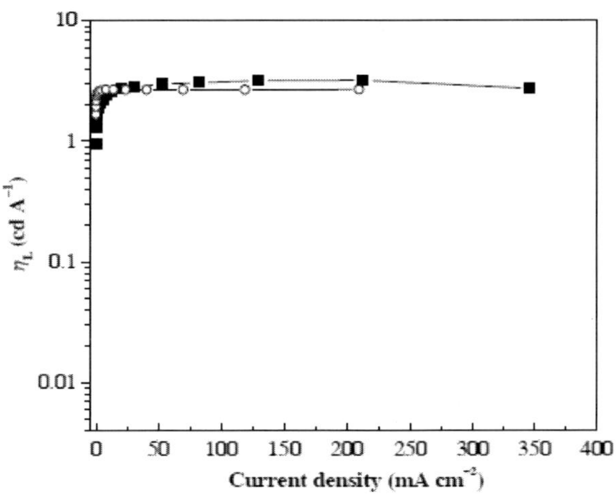

Figure 13. Luminance efficiency as a function of current density of device VIIc (!) and VIIIc (O).

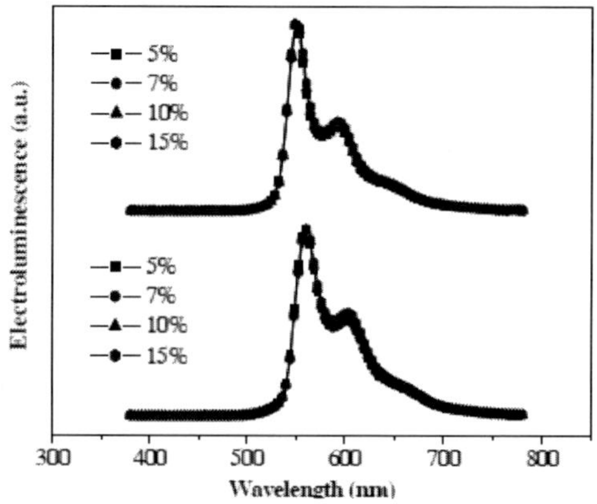

Figure 14. EL spectra of (a) 7 and (b) 9-doped solution-processed OLEDs at different dopant levels.

Table 5. Key performance data of electrophosphorescent OLEDs based on spin-coated emitting layer

Device	Dopant [wt.-%]	$V_{turn-on}$ [V]	L_{max} [cd m^{-2}] [a]	η_{ext} [%] [b]	max η_{ext} [%] [c]	max η_L [cd A¯1] [c]	max η_P [lm W¯1] [c]	λmax [nm] [d]
Va	**3** (5)	8.3	16020 (22)	3.91	5.11 (22)	4.56 (22)	0.65 (22)	624 (0.67, 0.33)
Vb	**3** (7)	7.9	10440 (22)	4.31	4.76 (21)	4.22 (22)	0.63 (21)	624 (0.67, 0.33)
V**c**	**3** (10)	5.5	13030 (23)	4.14	5.60 (7)	4.84 (7)	2.41 (9)	624 (0.67, 0.33)
V**d**	**3** (15)	4.6	11310 (21)	3.78	5.19 (7)	4.43 (7)	1.98 (7)	624 (0.67, 0.33)
VI**a**	**4** (5)	8.6	9961 (18)	6.74	7.65 (19)	6.18 (18)	1.07 (18)	624 (0.68, 0.32)
VI**b**	**4** (7)	9.5	16620 (21)	3.70	5.26 (21)	4.28 (21)	0.64 (21)	624 (0.68, 0.32)
VIc	**4** (10)	7.7	4629 (22)	3.27	3.48 (22)	2.75 (22)	0.75 (10)	624 (0.68, 0.32)
VI**d**	**4** (15)	6.4	7919 (22)	3.41	4.44 (7)	3.47 (8)	1.52 (7)	624 (0.68, 0.32)
VIIa	**5** (3)	8.8	17550 (20)	4.85	6.57 (20)	4.66 (18)	0.81 (18)	624 (0.68, 0.33)
VII**b**	**5** (5)	8.7	10770 (21)	5.34	7.18 (21)	4.65 (19)	0.76 (19)	624 (0.68, 0.32)
VIIc	**5** (10)	7.6	9357 (23)	4.05	4.17 (22)	3.18 (22)	0.51 (15)	624 (0.68, 0.32)
VII**d**	**5** (15)	5.9	7779 (21)	2.98	3.09 (20)	2.28 (20)	0.89 (6)	624 (0.68, 0.32)
VIII**a**	**6** (3)	8.8	10800 (20)	4.46	6.24 (20)	3.80 (19)	0.63 (15)	624 (0.68, 0.32)
VIIIb	**6** (5)	7.9	11540 (22)	4.27	5.64 (22)	3.37 (21)	0.63 (13)	624 (0.68, 0.32)
VIII**c**	**6** (10)	7.6	5576 (21)	3.54	3.56 (21)	2.66 (21)	0.67 (10)	624 (0.68, 0.32)
VIIId	**6** (15)	6.3	5041 (21)	2.83	3.03 (10)	2.17 (10)	0.87 (7)	624 (0.68, 0.32)
IXa	**7** (3)	6.7	21077 (16)	6.27	6.60 (15)	22.47 (15)	5.12 (13)	548 (0.44, 0.54)
IXb	**7** (5)	6.7	19080 (16)	6.68	7.18 (15)	24.22 (15)	6.76 (8)	548 (0.45, 0.54)
IXc	**7** (10)	6.6	15252 (17)	4.26	4.49 (9)	14.72 (16)	5.89 (8)	548 (0.45, 0.54)
IXd	**7** (15)	5.8	17708 (17)	3.98	4.06 (16)	13.94 (16)	4.65 (6)	548 (0.45, 0.53)
Xa	**8** (5)	7.0	26220 (18)	4.85	5.26 (15)	19.06 (15)	4.12 (17)	560 (0.48, 0.50)
Xb	**8** (10)	6.9	27090 (19)	5.2	5.60 (15)	17.51 (14)	7.99 (10)	560 (0.49, 0.50)
Xc	**8** (15)	6.7	21570 (18)	4.85	9.58 (9)	29.77 (9)	13.35 (9)	560 (0.49, 0.50)
XIa	**9** (5)	7.8	27680 (19)	3.49	3.83 (14)	11.90 (14)	3.45 (15)	560 (0.49, 0.50)
XIb	**9** (7)	7.8	21110 (18)	3.27	4.14 (13)	14.17 (14)	4.67 (9)	560 (0.49, 0.50)
XIc	**9** (10)	6.8	27030 (17)	3.64	5.54 (9)	17.23 (10)	6.76 (10)	560 (0.49, 0.50)
XId	**9** (15)	6.8	19540 (18)	2.77	3.24 (9)	10.06 (10)	3.81 (9)	560 (0.49, 0.49)
XIIa	**10** (5)	6.9	23790 (19)	3.37	3.70 (14)	10.51 (14)	2.79 (10)	568 (0.52, 0.47)
XIIb	**10** (10)	7.8	23790 (21)	3.53	3.94 (15)	11.19 (9)	2.69 (10)	568 (0.52, 0.47)
XIIc	**10** (15)	8.9	22040 (22)	3.07	3.44 (15)	9.78 (15)	2.49 (12)	568 (0.51, 0.47)

[a] Maximum luminance. Values in parentheses are the voltages at which the maximum values were obtained.

[b] Values collected at 100 mA/cm^2.

[c] Values in parentheses are the voltages at which the maximum values were obtained.

[d] CIE coordinates.

[x, y] in parentheses.

Although red, green and blue emitters of high color purity are required for full color display applications, there is still a great demand for emitters that give a bright color, such as orange for multiple color displays. Furthermore, there is a flurry of interest in employing orange emitters (generally regarded as a mixture of red and green colors) in combination with blue emitters to achieve a white light emission with simpler OLED architectures. [33] In the literature, only a rather limited number of orange electrophosphors for OLEDs are reported where almost all of them were fabricated by the vacuum deposition approach. [34] Orange-emitting active layers obtained by spin-coating method are very scarce to date.

Multilayer OLEDs using 7–10 as orange dyes in the doped emission layer have been fabricated (devices IX–XII in Figure 6) using the spin-coating method. Again, the typical HTL based on some classical triarylamines can be omitted in these devices. All of them possess very good film-forming properties and the highly amorphous nature of these phosphors favors a good chemical compatibility with the CBP host. Pertinent device performance data are collected in table 5. Figure 14 shows the EL spectra of 7 and 9 at different doping levels without any unwanted emission from CBP or Alq_3 even at high current density, which is a good indication of an effective energy transfer from the host exciton to the phosphor upon electrical excitation and the effective hole-blocking effect of BCP. The EL emission peaks are virtually independent of the driving voltage and do not show any concentration dependence. Each of the devices exhibits a strong orange EL peak with turn-on voltages ($V_{turn-on}$) of 5.8–7.0 V and 6.8–8.9 V for the homo- and heteroleptic species, respectively.

All of the new phosphors display excellent OLED performance since the end-capping carbazole unit may enhance charge trapping. Figure 15 presents current density and luminance versus bias voltage (*I–V–L*) curves of the OLEDs for 7 and 9 at 10 wt.-% concentrations. The best device performance is achieved for 7 at the doping level of 5 wt.-% in which a maximum luminance of 19080 cd m at 16 V and a peak η_{ext} of 7.18 %, corresponding to a peak η_L of 24.22 cd A^{-1} and a peak η_p of 6.76 lm W^{-1} were obtained (Figure 16). The performance of 8-doped devices is also very remarkable with the peak efficiencies at η_{ext} = 9.58 %, η_L = 29.77 cd A^{-1} and η_p = 13.35 lm W^{-1} at 15 wt.-% concentration (Figure 16). Remarkably, the efficiency roll-off is not significant when the current density increases from 20 to 100 mA cm^{-2}. The EL efficiencies of these orange OLEDs are some of the highest reported for solution-processed devices and are even comparable to many of the best vacuum-evaporated orange-emitting OLEDs.

Both devices at the 10 wt.-% doping level show peak η_{ext} = 5.54 %, η_L = 17.23 cd A^{-1} and η_p = 6.67 lm W^{-1} for 9 and η_{ext} = 3.94 %, η_L = 11.19 cd A^{-1} and η_p = 2.69 lm W^{-1} for 10 (Figure 17). Similar to the case for 7, owing to the short lifetimes of 8–10, the device efficiencies for 8–10 witnessed a slow decay with increasing driving voltage and current density, demonstrating their relatively high device stability. Our results reveal that the overall efficiencies are higher in the homoleptic complexes than their heteroleptic ones (i.e. 7 > 9 and 8 > 10). Such performance order may be attributed to a higher PL triplet yield of the former as compared to the latter series. From the *I–V–L* characteristics for the 7- and 9-doped devices, the higher device conductivity for the heteroleptic species in general (e.g. 36.4 and 93.5 mA cm^{-2} for 7 and 9, respectively, at 15 V for 10 % doping level) indicates that the holes and electrons are passing into the active layer without effectively forming excitons which is partly attributable to the lower trapping efficiency. A lower exciton population is then generated inside the active layer, yielding a lower η_{ext} for 9 relative to 7. The better EL performance of 7 may also be attributed to its blue-shifted emission, in which the relatively large energy gap is less perturbed by nonradiative processes according to the energy-gap law. [35] Furthermore, probably due to a closer match of the HOMO of 7 and 8 with that for PEDOT (HOMO –5.0 eV), we observed that the devices doped with the homoleptic guests show higher η_{ext} and η_L values than those based on the heteroleptic congeners.

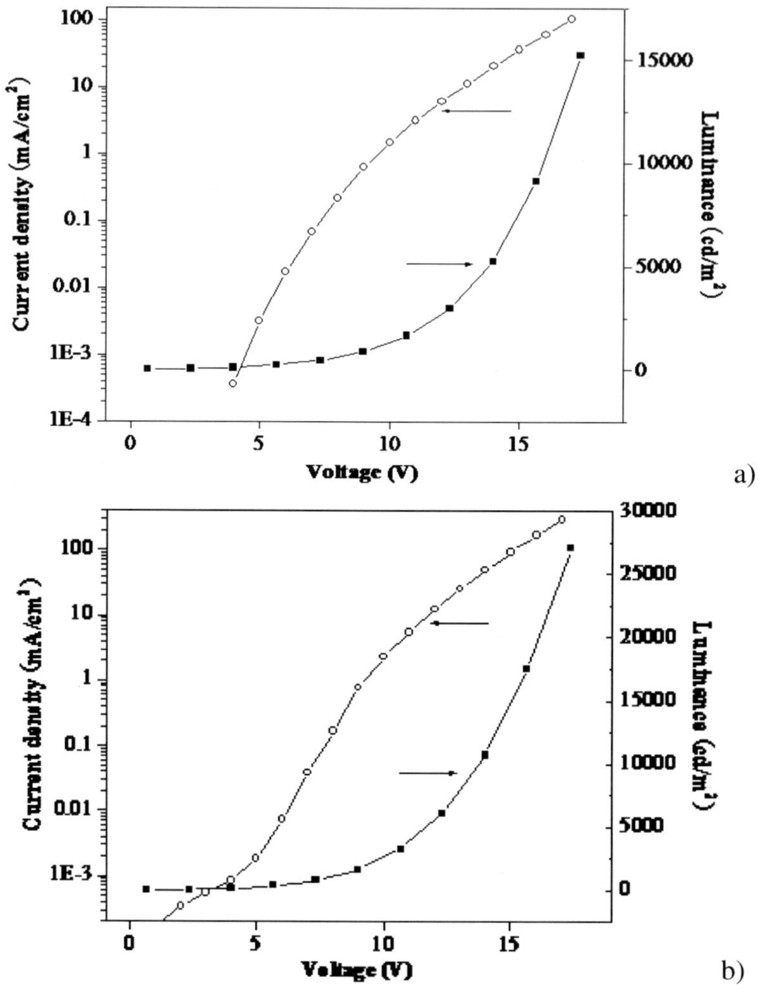

Figure 15. Current–voltage–luminance (*I–V–L*) characteristics of the electrophosphorescent OLED devices with 10 wt.-% dopant levels of (a) 7 and (b) 9.

Conclusion and Outlook

In summary, we have developed some triplet-harvesting carbazolyl-substituted multi-component iridium homoleptic and heteroleptic complexes and this strategy can offer advantages in terms of lowering the first IP, facilitating HT and enhancing EL efficiencies as compared to *fac*-[Ir(ppy)$_3$]. While the HOMO levels are raised after inclusion of the 9-arylcarbazole module, the emissive energies for the green and red colors are relatively invariant of the X group on the 9-aryl ring. The performance parameters of the green OLED devices are impressive at about 12 % and 38 cd A^{-1}. Red OLED devices based on these materials show saturated red electrophosphorescence with impressive CIE color coordinates of (0.67, 0.33) to (0.68, 0.32). The external quantum efficiency as high as 12 % can be achieved for vacuum-evaporated devices. On the other hand, the enhanced HT properties of the complexes have been demonstrated for devices based on spin-coated phosphorescent layer without the need for an additional HTL and we note that the end-capping carbazole unit may enhance charge trapping. The results show that the carbazole ring can improve the charge-balance properties in the EL process, rendering these materials bifunctional in nature. The efficiencies of our devices with CIE (0.68, 0.32) are some of the highest ever reported for pure red vacuum-deposited and spin-coated devices with a similar CIE color. More importantly, these new materials should provide a good venture for the design of metallophosphors with optimized OLED efficiency/color purity trade-offs essential for pure red light, which still remains as a great challenge for red light emitters.

We have also elegantly combined the carbazole-fluorene hybrid fluorophore with the phosphorescent heavy metal core to yield some solution-processible homoleptic and heteroleptic orange-emitting iridium(III) cyclometalates. The advance of new orange phosphors in this way can provide a practical basis for the future design of white light illumination sources by blending the orange dyes with a blue fluorescent or phosphorescent emitter. In essence, these multi-component systems have a great potential to excel in accomplishing WOLEDs through various device architectures. Given the ease of synthesis and performance advantage in different spectral color region (namely, green, orange and red) for these carbazole-based metallophosphors, extension of the system to other emission colors would warrant further examination.

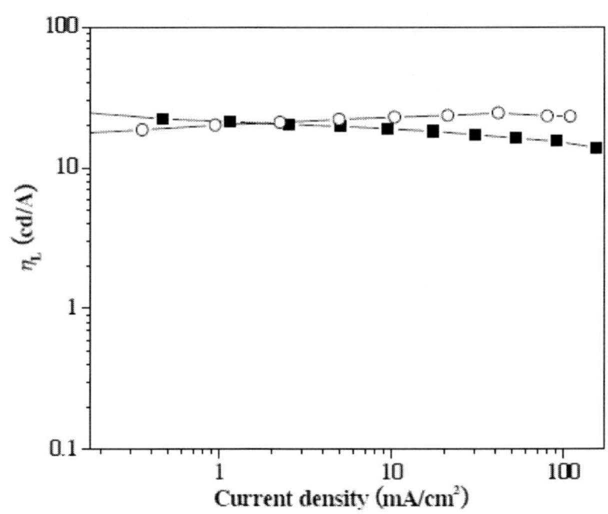

Figure 16. Luminance efficiency as a function of current density for OLED devices using (a) 5 wt.-% 7 (■) and (b) 15 wt.-% 8 (O).

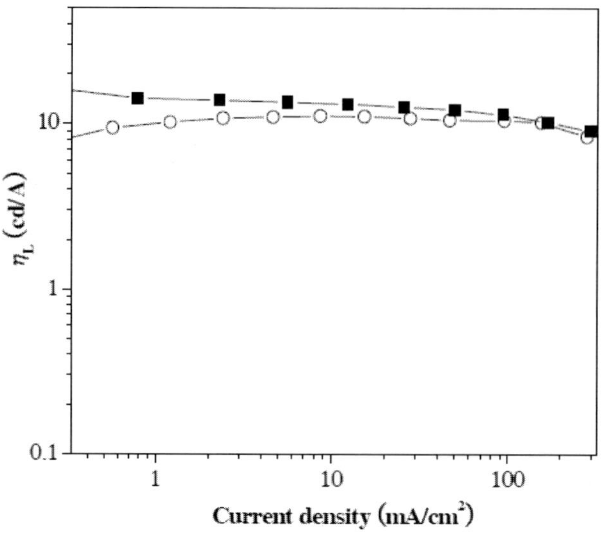

Figure 17. Luminance efficiency as a function of current density for OLED devices using (a) 10 wt.-% 9 (■) and (b) 10 wt.-% 10 (○).

ACKNOWLEDGMENT

This work was supported by CERG Grants from the Research Grants Council of the Hong Kong SAR, P.R. China (Project No. HKBU 202106) and the Hong Kong Baptist University. W.-Y. W. would also like to thank his postgraduate students, postdoctoral associates and collaborators whose names appear in the references.

REFERENCES

[1] (a) Baldo, M. A.; Lamansky, S.; Burrows, P. E.; Thompson, M. E.; Forrest, S. R. *Appl. Phys. Lett.* 1999, 75, 4; (b) Baldo, M. A.; Thompson, M. E.; Forrest, S. R. *Nature* 2000, 403, 750.

[2] (a) Gong, X.; Robinson, M. R.; Ostrowki, J. C.; Moses, D.; Bazan, G. C.; Heeger, A. J. *Adv. Mater.* 2002, 14, 581; (b) Köhler, A.; Wilson, J. S.; Friend, R. H. *Adv. Mater.* 2002, 14, 701; (c) Veinot, J. G. C.; Marks, T. J. *Acc. Chem. Res.* 2005, 38, 632; (d) Forrest, S. R. *Org. Electron.* 2003, 4, 45; (e) Chou, P. T.; Chi, Y. *Chem. Soc. Rev.* 2007, 36, 1421.

[3] (a) Chou, P. T.; Chi, Y. *Chem. Eur. J.* 2007, 13, 380; (b) Yersin, H. *Top. Curr. Chem.* 2004, 241, 1; (c) Langeveld, B. M. W.; Schubert, U. S. *Adv. Mater.* 2005, 17, 1109; (d) Baldo, M. A.; Thompson, M. E.; Forrest, S. R. *Pure Appl. Chem.* 1999, 71, 2095; (e) Evans, R. C.; Douglas, P.; Winscom, C. J. *Coord. Chem. Rev.* 2006, 250, 2093; (f) Burn, P. L.; Lo, S.-C.; Samuel, I. D. W. *Adv. Mater.* 2007, 19, 1675; (g) Lowry, M. S.; Bernhard, S. *Chem. Eur. J.* 2006, 12, 7979.

[4] (a) Adachi, C.; Baldo, M. A.; Forrest, S. R.; Thompson, M. E. *Appl. Phys. Lett.* 2000, 77, 904; (b) Ikai, M.; Tokito, S.; Sakamoto, Y.; Suzuki, T.; Taga, Y. *Appl. Phys. Lett.* 2001, 79, 156; (c) Adachi, C.; Baldo, M. A.; Thompson, M. E.; Forrest, S. R. *J. Appl. Phys.* 2001, 90, 5048; (d) Baldo, M. A.; O'Brien, D. F.; You, Y.; Shoustikov, A.; Sibley, S.; Thompson, M. E.; Forrest, S. R. *Nature* 1998, 395, 151.

[5] (a) Lamansky, S.; Djurovich, P.; Murphy, D.; Abdel-Razzaq, F.; Lee, H. E.; Adachi, C.; Burrows, P. E.; Forrest, S. R.; Thompson, M. E. *J. Am. Chem. Soc.* 2001, 123, 4304; (b) Kido, J.; Shionoya, H.; Nagai, K. *Appl. Phys. Lett.* 1995, 67, 2281; (c) Granstrom, M.; Inganas, O. *Appl. Phys. Lett.* 1996, 68, 147; (d) Jiang, X.; Zhang, Z.; Zhao, W.; Zhu, W.; Zhang, B.; Zu, S. *J. Phys. D.: Appl. Phys.* 2000, 33, 473; (e) D'Andrade, B. W.; Thompson, M. E.; Forrest, S. R. *Adv. Mater.* 2002, 14, 147; (f) Wong, K. M. C.; Zhu, X.; Hung, L. L.; Zhu, N.; Yam, V. W. W.; Kwok, H. S. *Chem. Commun.* 2005, 2906; (g) Chen, Y. L.; Li, S. W.; Cheng, Y. M.; Pu, S. C.; Yeh, Y. S.; Chou, P.-T. *Chem. Phys. Chem.* 2005, 6, 2012; (h) Zhou, G. J.; Wong, W.-Y.; Yao, B.; Xie, Z.; Wang, L. *Angew. Chem. Int. Ed.* 2007, 46, 1149; (i) Nazeeruddin, M. K.; Wegh, R. T.; Zhou, Z.; Klein, C.; Wang, Q.; De Angelis, F.; Fantacci, S.; Gratzel, M. *Inorg. Chem.* 2006, 45, 9245; (j) Zhou, G.-J.; Ho, C.-L.; Wong, W.-Y.; Wang, Q.; Ma, D.-G.; Wang, L.-X.; Lin, Z.-Y.; Marder, T. B.; Beeby, A. *Adv. Funct. Mater.* 2008, 18, 499.

[6] (a) Lundin, N. J.; Blackman, A. G.; Gordon, K. C.; Officer, D. L. *Angew. Chem. Int.* Ed. 2006, 45, 2582; (b) Chisholm, M. H.; Epstein, A. J.; Gallucci, J. C.; Feil, F.; Pirkle, W. *Angew. Chem. Int. Ed.* 2005, 44, 6537; (c) Gong, X.; Ng, P. K.; Chan, W. K. *Adv. Mater.* 1998, 10, 1337; (d) Wong, W.-Y.; He, Z.; So, S. K.; Tong, K. L.; Lin, Z. *Organometallics.* 2005, 24, 4079; (e) Liu, Z.; Guan, M.; Bian, Z.; Nie, D.; Gong, Z.; Li, Z.; Huang, C. *Adv. Funct. Mater.* 2006, 16, 1441; (f) He, Z.; Wong, W.-Y.; Yu, X.; Kwok, H. S.; Lin, Z. *Inorg. Chem.* 2006, 45, 10922.

[7] (a) Pai, D. M.; Yanus, J. F.; Stolka, M. *J. Phys. Chem.* 1984, 88, 4714; (b) Kanai, H.; Ichinosawa, S.; Sato, Y. *Synth. Met.* 1997, 91, 195.

[8] (a) Yeh, S.-J.; Wu, M.-F.; Chen, C.-T.; Song, Y.-H.; Chi, Y.; Ho, M.-H.; Hsu, S.-F.; Chen, C. H. *Adv. Mater.* 2005, 17, 285; (b) Brunner, K.; van Dijken, A.; Börner, H.; Bastiaansen, J. J. A. M.; Kiggen, N. M. M.; Langeveld, B. M. W. *J. Am. Chem. Soc.* 2001, 123, 9404; (c) Strohriegl, P.; Grazulevicius, J. V. *Adv. Mater.* 2003, 15, 445.

[9] Wong, W.-Y.; Ho, C.-L.; Gao, Z.-Q.; Mi, B.-X.; Chen, C. H.; Cheah, K. W.; Lin, Z. *Angew. Chem. Int.* Ed. 2006, 45, 7800; (b) Ho, C.-L.; Wong, W.-Y.; Gao, Z.-Q.; Chen, C. H.; Cheah, K.-W.; Yao, B.; Xie, Z.; Ma, D.; Wang, L.; Yu, X.-M.; Kwok, H.-S.; Lin, Z. *Adv. Funct. Mater.* 2008, 18, 319; (c) Ho, C.-L.; Wong, W.-Y.; Zhou, G.-J.; Yao, B.; Xie, Z.; Wang, L. *Adv. Funct. Mater.* 2007, 17, 2925.

[10] Shen, L.; Chen, Z.; Zhao, Q.; Li, F. Y.; Yi, T.; Cao, Y.; Huang, C. H. *Inorg. Chem. Commun.* 2006, 9, 620.

[11] Tamayo, A. B.; Garon, S.; Sajoto, T.; Djurovich, P. I.; Tsyba, I. M.; Bau, R.; Thompson, M. E. *Inorg. Chem.* 2005, 44, 8723.

[12] King, K. A.; Spellane, P. J.; Watts, R.-J. *J. Am. Chem. Soc.* 1985, 107, 1431.

[13] Hay, P. J. *J. Phys. Chem.* A. 2002, 106, 1634.

[14] Wilde, A. P.; King, K. A.; Watts, R. J. *J. Phys. Chem.* 1991, 95, 629; b) Sprouse, S.; King, K. A.; Spellane, P. J.; Watts, R. J. *J. Am. Chem. Soc.* 1984, 106, 6647; c) Columbo, M. C.; Hauser, A.; Güdel, H. U. *Top. Curr. Chem.* 1994, 171, 143.

[15] (a) Wang, Y.; Herron, N.; Grushin, V. V.; LeCloux, D.; Petrov, V. *Appl. Phys. Lett.* 2001, 79, 449; (b) Garces, F. O.; King, K. A.; Watts, R. J. *Inorg. Chem.* 1988, 27, 3464.

[16] Spellane, P. J.; Watts, R. J. *Inorg. Chem.* 1993, 32, 5633.

[17] Tsuboyama, A.; Iwawaki, H.; Furugori, M.; Mukaide, T.; Kamatani, J.; Igawa, S.; Moriyama, T.; Miura, S.; Takiguchi, T.; Okada, S.; Hoshino, M.; Ueno, K. *J. Am. Chem. Soc.* 2003, 125, 12971.

[18] (a) Coppc, P.; Plummer, E. A.; De Cola, D. *Chem. Commun.* 2004, 1774; (b) Ostrowski, J. C.; Robinson, M. R.; Heeger, A. J.; Bazan, G. C. *Chem. Commun.* 2002, 784.

[19] (a) Baldo, M. A.; Adachi, C.; Forrest, S. R. *Phys. Rev.* B 2000, 62, 10967. (b) Namdas, E. B.; Ruseckas, A.; Samuel, I. D. W.; Lo, S. C.; Burn, P. L. *Appl. Phys. Lett.* 2005, 86, 091104.

[20] Sajoto, T.; Djurovich, P. I.; Tamayo, A.; Yousufuddin, M.; Bau, R.; Thompson, M. E. *Inorg. Chem.* 2005, 44, 7992.

[21] (a) Thelakkat, M.; Schmidt, H.-W. *Adv. Mater.* 1998, 10, 219; (b) Ashraf, R. S.; Shahid, M.; Klemm, E.; Al-Ibrahim, M.; Sensfuss, S. *Macromol. Rapid Commun.* 2006, 27, 1454.

[22] Li, C. L.; Su, Y. J.; Tao, Y. T.; Chou, P. T.; Chien, C. H.; Cheng, C. C.; Liu, R. S. *Adv. Funct. Mater.* 2005, 15, 387.

[23] van Dijken, A.; Bastiaansen, J. J. A. M.; Kiggen, N. M. M.; Langeveld, B. M. W.; Rothe, C.; Monkman, A.; Bach, I.; Stossel, P.; Brunner, K. *J. Am. Chem. Soc.* 2004, 126, 7718.

[24] (a) Copper, A. D.; Cok, R. S.; Feldman, R. D. *Proc. S. P. I. E. – Int. Soc. Opt. Eng.* 2001, 4105, 18; (b) C.-T. Chen, *Chem. Mater.* 2004, 16, 4389.

[25] Okada, S.; Okinaka, K.; Iwawaki, H.; Furugori, M.; Hashimoto, M.; Mukaide, T.; Kamatani, J.; Igawa, S.; Tsuboyama, A.; Takiguchi, T.; Ueno, K. *Dalton. Trans.* 2005, 1583.

[26] (a) D'Andrade, B. W.; Baldo, M. A.; Adachi, C.; Brooks, J.; Thompson, M. E.; Forrest, S. R. *Appl. Phys. Lett.* 2001, 79, 1045; (b) Adachi, C.; Kwong, R. C.; Djurovich, P.; Adamovich, V.; Baldo, M. A.; Thompson, M. E.; Forrest, S. R. *Appl. Phys. Lett.* 2001, 79, 2082.

[27] Jiang, C.; Yang, W.; Peng, J.; Xiao, S.; Cao, Y. *Adv. Mater.* 2004, 16, 537.

[28] Anthopoulos, T. D.; Frampton, M. I.; Namdas, E. B.; Burn, P. L.; Samuel, I. D. W. *Adv. Mater.* 2004, 16, 557.

[29] Rayabarapu, D. K.; Paulose, B. M. J. S.; Duan, J.-P.; Cheng, C.-H. *Adv. Mater.* 2005, 17, 349.

[30] Fujii, H.; Sakurai, H.; Tani, K.; Wakisaka, K.; Hirao, T. *I. E. I. C. E. Electronics Express*. 2006, 2, 260.

[31] Su, Y. J.; Huang, H. L.; Li, C. L.; Chien, C. H.; Tao, Y. T.; Chou, P. T.; Datta, S.; Liu, R. S. *Adv. Mater.* 2003, 15, 884.

[32] Kanno, H.; Sun, Y.; Forrest, S. R. *Appl. Phys. Lett.* 2005, 86, 263502.

[33] (a) Misra, A.; Kumar, P.; Kamalasanan, M. N.; Chandra, S. *Semicond. Sci. Technol.* 2006, 21, R35; (b) Yu, X.-M.; Kwok, H.-S.; Wong, W.-Y.; Zhou, G.-J. *Chem. Mater.* 2006, 18, 5097; (c) D'Andrade, B. W.; Forrest, S. R. *Adv. Mater.* 2004, 16, 1585; (d) Fuhrmann, T.; Salbeck, J. *M. R. S. Bull.* 2003, 28, 354.

[34] (a) Shoustikov, A.; You, Y.; Burrows, P. E.; Thompson, M. E.; Forrest, S. R. *Synth. Met.* 1997, 91, 217; (b) Laskar, I. R.; Hsu, S. F.; Chen, T. M. *Polyhedron.* 2005, 24, 881; (c) Wong, W.-Y.; Zhou, G.-J.; Yu, X.-M.; Kwok, H.-S.; Tang, B.-Z. *Adv. Funct. Mater.* 2006, 16, 838; (d) Wong, W.-Y.; Zhou, G.-J.; Yu, X.-M.; Kwok, H.-S.; Lin, Z. *Adv. Funct. Mater.* 2007, 17, 315.

[35] (a) Cummings, S. D.; Eisenberg, R. *J. Am. Chem. Soc.* 1996, 118, 1949; (b) Wilson, J. S.; Chawdhury, N.; Al-Mandhary, M. R. A.; Younus, M.; Khan, M. S.; Raithby, P. R.; Köhler, A.; Friend, R. H. *J. Am. Chem. Soc.* 2001, 123, 9412.

INDEX

#

A

B

C

I

J

K

L

M

N

O

S

T

U

V